SpringerBriefs in Applied Sciences and Technology

SpringerBriefs present concise summaries of cutting-edge research and practical applications across a wide spectrum of fields. Featuring compact volumes of 50 to 125 pages, the series covers a range of content from professional to academic.

Typical publications can be:

- A timely report of state-of-the art methods
- An introduction to or a manual for the application of mathematical or computer techniques
- A bridge between new research results, as published in journal articles
- A snapshot of a hot or emerging topic
- An in-depth case study
- A presentation of core concepts that students must understand in order to make independent contributions

SpringerBriefs are characterized by fast, global electronic dissemination, standard publishing contracts, standardized manuscript preparation and formatting guidelines, and expedited production schedules.

On the one hand, **SpringerBriefs in Applied Sciences and Technology** are devoted to the publication of fundamentals and applications within the different classical engineering disciplines as well as in interdisciplinary fields that recently emerged between these areas. On the other hand, as the boundary separating fundamental research and applied technology is more and more dissolving, this series is particularly open to trans-disciplinary topics between fundamental science and engineering.

Indexed by EI-Compendex, SCOPUS and Springerlink.

More information about this series at http://www.springer.com/series/8884

David Bienvenido-Huertas · Carlos Rubio-Bellido

Optimization of the Characterization of the Thermal Properties of the Building Envelope

Analysis of the Characterization of the Façades using Artificial Intelligence

Springer

David Bienvenido-Huertas
Higher Technical School of Building Engineering
University of Seville
Seville, Spain

Carlos Rubio-Bellido
Higher Technical School of Building Engineering
University of Seville
Seville, Spain

ISSN 2191-530X ISSN 2191-5318 (electronic)
SpringerBriefs in Applied Sciences and Technology
ISBN 978-3-030-63628-9 ISBN 978-3-030-63629-6 (eBook)
https://doi.org/10.1007/978-3-030-63629-6

This Springer imprint is published by the registered company Springer Nature Switzerland AG
The registered company address is: Gewerbestrasse 11, 6330 Cham, Switzerland

Contents

Chapter 1
The Influence of the Envelope Thermal Properties on Building Energy Performance

1.1 Low Carbon Economy Goals

From ancient times, human beings have taken advantage of natural resources as much as possible for their own benefit, although that advantage is limited in order to guarantee the environment sustainability. After the Industrial Revolution, however, the balance between human beings and nature was broken, mainly due to the need of society to meet its growing production by excessively exploiting the natural resources of the biosphere [1]. The interaction of society with the environment is negatively influencing nature, leading to climate change, the acidification of oceans, and the extinction of species [2]. In addition, social and demographic aspects are also affected, such as environmental refugees. For 40 years, the capacity of earth resources has been exceeded [2], thus reflecting the unsustainability of the society's life system from the twenty-first century. For this reason, the major economic powers are increasingly worried about the excessive dependence of fossil fuels as these resources are imported from countries with an unstable political activity. This fact implies to be subjected to potential economic crisis, such as the oil crisis of 1973 and 1979.

In addition, the possible climate evolution scenarios throughout the twenty-first century are not encouraging. According to the Intergovernmental Panel on Climate Change [3], those scenarios (i.e. scenarios B1, A1T, B2, A1B, A2, and A1FI) claim that the temperature will increase between 1.1 °C and 6.4 °C by the end of the twenty-first century, and the sea level will increase between 18 and 59 cm in comparison with the values from the end of the twentieth century.

One of the main reasons is the greenhouse gases continuously emitted into the atmosphere. Many sectors have high percentages of greenhouse gas emissions. In this regard, the data included in the report entitled United in Science [4], published in the UN Climate Action Summit 2019, present a worrying trend: an annual increase of 1% in CO_2 emissions, a predominance of fossil fuel consumption (despite renewable energies are more and more used), and an increase in CO_2, CH_4, and N_2O of 146%,

D. Bienvenido-Huertas and C. Rubio-Bellido, *Optimization of the Characterization of the Thermal Properties of the Building Envelope*, SpringerBriefs in Applied Sciences and Technology, https://doi.org/10.1007/978-3-030-63629-6_1

275%, and 122%, respectively, in comparison with preindustrial periods (i.e. before 1750).

International bodies are therefore focused on setting goals for the world population to achieve a more sustainable development to protect the environment. An example of this new society's trend is the international congresses and conventions taken place in the last 25 years. These congresses and conventions led to treaties, such as the Kyoto Protocol and the Paris Agreement. The latter defined demanding goals to maintain climate conditions against the progressive temperature increase. One of the goals was maintaining the temperature increase below 2 °C in comparison with preindustrial periods. However, this is the worst scenario as the agreement urges to achieve an increase of only 1.5 °C.

Moreover, this agreement forces the 195 countries to develop strategies and policies focused on improving the efficiency and sustainability of productive processes. Thus, these countries established demanding goals to achieve a low-carbon economy by 2050.

The European Union's roadmap towards a low-carbon economy aims to reducing the greenhouse gas emissions produced by European countries [5]. The goal in the first stage was to reduce these emissions by 20% by 2020, but this goal is an intermediate stage within the final goal, that is, reducing greenhouse gas emissions between 80 and 90% by 2050. For this purpose, goals based on reducing greenhouse gas emissions are set in all major sectors, including the residential sector.

1.2 Building Energy Efficiency

Construction is among the most significant activities carried out by human beings as it is one of the most complex sectors that greatly impacts society. In addition, this sector is continuously developed. Users' activities in residential buildings are significant in construction as people spend most of their time in them, and consequently users' behaviour patterns could significantly affect the energy use [6, 7].

Approximately 40% of the total energy consumption from human activities is related to the building sector [8], which generates 38% of greenhouse gas emissions [9]. In 2010, the total energy consumption of the building sector was 23.7 PWh, and it is estimated that this consumption could reach 38.4 PWh by 2040 [10].

This energy consumption is mainly used for heating and air-conditioning systems, water warming, electrical household appliances, and lighting systems. However, several studies have emphasized that the main source of energy consumption in these buildings is the consumption generated by heating, ventilation, and air-conditioning (HVAC) systems [11, 12].

Although the energy consumption could vary due to factors related to building properties [13], socioeconomic factors could also influence the energy consumption of dwellings [14, 15]. Thus, the relationship between the building energy demand, users' behaviour patterns, and social and economic factors could significantly influence this type of consumption. This aspect becomes more important if the effect of

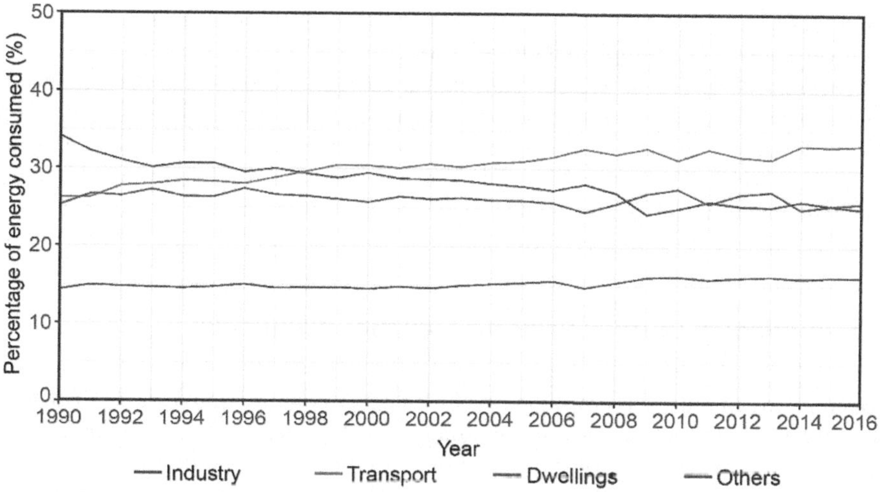

Fig. 1.1 Final energy consumption in Europe by sectors. Data used are those from the European environment agency [16]

dwelling performance is considered at a global scale [13] as minor variations could imply significant savings.

The deficient performance is reflected in the high energy consumption recorded in buildings in recent years. In the European Union, residential buildings represented 25.7% of the total primary energy consumption in 2016 [16]. Since 1990, the total energy consumption annually increases 1% with a peak of 2.5% in relation to the electrical demand [16] (Fig. 1.1), thus showing the progressive energy performance loss of the European building stock.

Most energy consumed in the use phase of residential buildings is from non-renewable resources, so the energy consumption in the existing building stock should be reduced. Within this context, the international community has intensified its efforts to reduce the increases of CO_2 emissions and the energy consumption related to the building sector.

Thus, the major framework to achieve a low-carbon residential sector is the state regulation on building energy efficiency. European buildings should reduce their greenhouse gas emissions by 90% by 2050 [5], so a larger number of nearly zero energy buildings (nZEB) is required. According to the Directive 2010/31/UE (European Union 2010), all state members should include in their state regulations the obligatory nature that new buildings or buildings to be restored are nZEB in the following dates:

- After 31 December 2018 in public buildings.
- After 31 December 2020 in all new buildings.

Nevertheless, the limitations for their application in warm climates is a challenge for the scientific and technological community [17]. Attia et al. [18] presented the difficulty of most countries in southern Europe to establish specific regulations focused on the correct implementation of nZEB.

These buildings are aimed at balancing the energy consumed and that produced, so designs that guarantee a null energy demand should be presented. As mentioned above, the consumption generated by HVAC systems is among the most significant types of energy consumption [11, 12]. This consumption is related to the building energy demand, which is also related to the envelope thermal properties.

1.3 Influence of the Envelope Thermal Properties

To reduce greenhouse gas emissions by 90%, energy conservation measures (ECMs) should be adopted in the existing buildings to reduce their energy demand. ECM is understood as any type of modification carried out in a building to improve its energy performance [19]. The envelope is closely related to the energy demand, so the envelope elements are generally the main element to deal with [20–23]. The effectiveness of applying ECMs to the building envelope has been widely studied, mainly analysing the energy improvement of buildings located in cold climate zones. Aksoy and Inalli [24] analysed the influence of passive design parameters, such as the form factor and orientation, in a building located in a cold region in Turkey. Similarly, Güçyeter and Günaydin [25] used 6 ECMs to assess the improvement of the envelope of an office building located in Turkey. Invidiata et al. [26] analysed the influence of 6 ECMs in a residential building located in the north of Italy to select the best option to improve the sustainability of the building. In a study at a greater scale, Qian et al. [27] assessed the saving achieved at a national level by combining various ECMs in commercial buildings.

These modifications in the envelope are important because the envelope thermal properties influence the building energy performance. In this regard, the envelope elements contribute to building heat losses or gains [20–23], thus leading to a greater solar control, the use of greater thicknesses of insulating materials, and the reduction of air filtrations, among others, which implies that buildings have a better energy performance. However, dynamic thermal properties are usually underestimated when designing envelopes [28]. Most analyses performed in the existing buildings are based on the stationary thermal transmittance. This variable, despite its importance to characterize the building energy performance, maybe does not consider some aspects related to the energy performance, for instance, the thermal inertia [29]. In this regard, considering both the stationary thermal transmittance and the thermal mass varies the building energy performance [30], with a greater effect in warm zones than in cold zones [31]. Other thermal properties, such as the periodic thermal properties developed in ISO 13786 [32], allow the expected energy performance of the building to be greater controlled. Some authors have emphasized the importance of these thermal variables. The first studies by Refs. [33–36] reflected the importance of controlling the thermal inertia, the periodic thermal transmittance, and the thermal admittance in the energy performance. Then, Stazi et al. [37] proved the importance of controlling the indoor thermal inertia and the decrement factor.

However, most of these studies analysed a small sample of walls (the number of walls was less than 20 in most cases). In a recent study by Bienvenido-Huertas et al. [38], a sample of 2413 walls was analysed in a virtual prototype of case study (Fig. 1.2), which was placed in both a cold and warm climate zone. This analysis considered the thermal transmittance of ISO 6946 [39] and the periodic thermal variables of ISO 13786 [32]. The periodic thermal variables were the periodic thermal transmittance, the time shift periodic thermal transmittance, the decrement factor, the time shift internal side, the external thermal admittance, and the time shift external side. The results were obtained through energy simulations in EnergyPlus, and the correlations between the energy demand and the thermal variables were analysed.

The results obtained from the relationship between the total energy demand of the case study and the stationary thermal variable are presented in Fig. 1.3. In the stationary thermal transmittance of ISO 6946, there was a linear tendency with the energy demand values. Thus, low values of stationary thermal transmittance implied

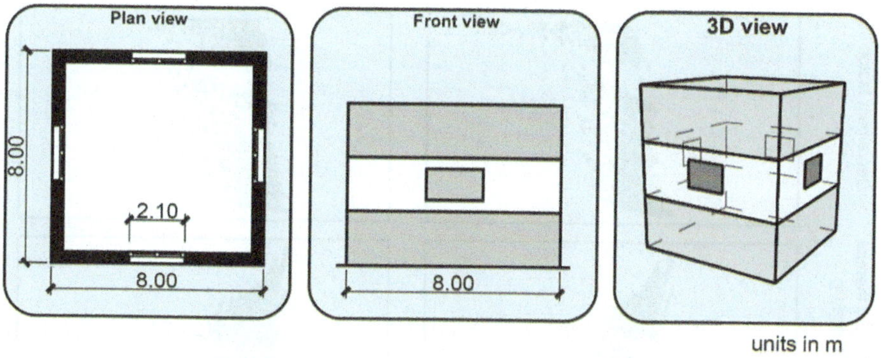

Fig. 1.2 Theoretical case study used by Bienvenido-Huertas et al. [38]

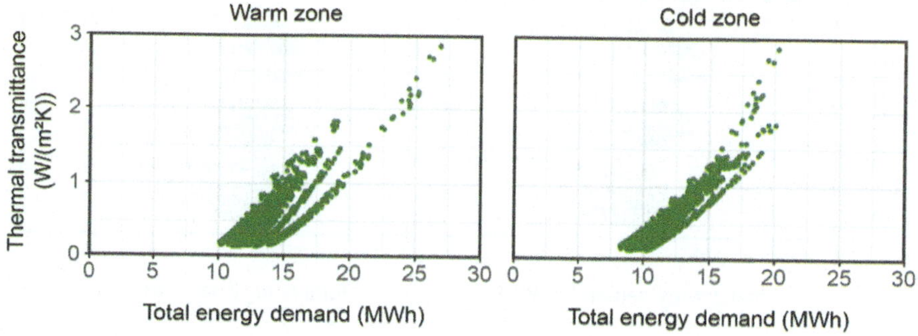

Fig. 1.3 Relationship between the thermal transmittance and the total energy demand in the two climate zones

low values of the energy demand of the building. However, the great oscillation of a same stationary thermal transmittance value in the energy demand (with variations up to 2 MWh in the warm zone and 1.6 MWh in the cold zone) forced to consider other variables to reduce the building energy demand.

Regarding the periodic thermal variables, there were various tendencies according to the variable analysed (Figs. 1.4 and 1.5). In these cases, the periodic thermal transmittance, the decrement factor, and the time shift obtained low values of energy demand with the optimal values of each variable (i.e. the lowest value of the periodic

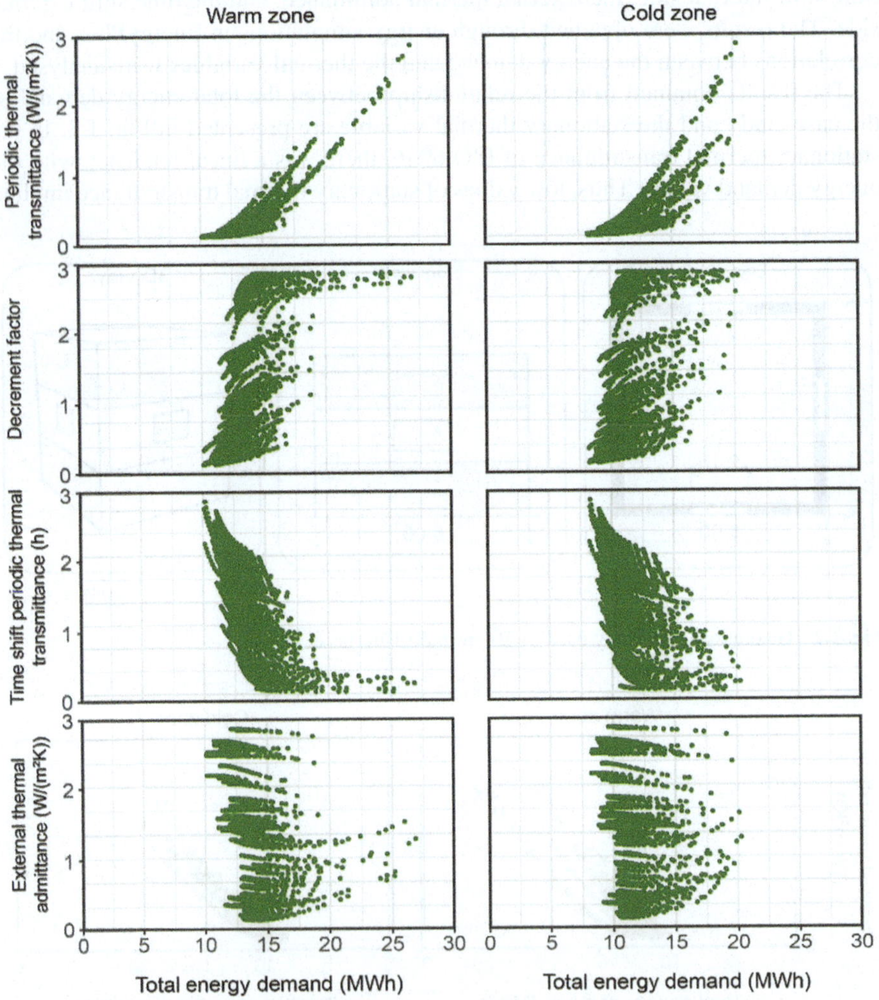

Fig. 1.4 Relationship between the periodic thermal variables (the periodic thermal transmittance, the time shift periodic thermal transmittance, the decrement factor, and the external thermal admittance) and the total energy demand in the two climate zones analysed

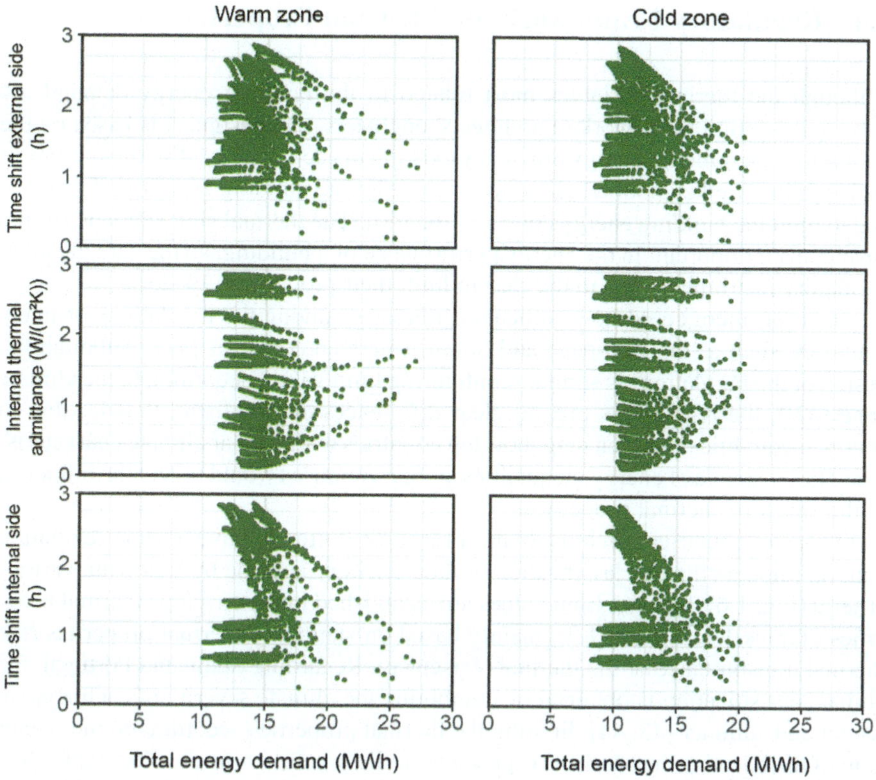

Fig. 1.5 Relationship between the periodic thermal variables (the time shift external side, the internal thermal admittance, and the time shift internal side) and the total energy demand in the two climate zones analysed

thermal transmittance or the decrement factor, and the highest value of the time shift periodic thermal transmittance). These periodic variables therefore control whether the design for the ECM is going to achieve a low total energy demand, while the other periodic variables, although they can be important for the design and estimation of the building energy performance, do not show a clear relationship between their values and the building energy performance. Thus, a correct control of the stationary thermal transmittance, the periodic thermal transmittance, the decrement factor, and the time shift would guarantee that the building energy performance obtained through the ECM is appropriate.

1.4 Regulatory Framework for Thermal Properties

Although the thermal variables most related to the building energy demand are known, the action framework of architects, engineers, and designers is based on the regulations on building energy efficiency existing in each country. The regulations of each country are key both to guarantee an appropriate building energy performance and to limit the building energy consumption [40]. The thermal characteristics of the envelope are important in the energy performance of a building, so they are the main regulatory instruments due to the ease to limit their properties [20–23].

However, there could be significant differences among the regulations of many countries, such as the European and South American countries [41]. Although the design of the regulations is similar (configuring limit values according to the climate zone [42]), the regulations present clear differences among them, thus leading to energy inequalities between neighbouring countries with similar climate characteristics. The existence of energy inequalities among countries would reflect the influence of the envelope thermal properties.

European countries, such as Spain, France or Portugal, have a same regulation pattern. First, a climate classification of their territory is made by following various criteria (Fig. 1.6). Second, limit values are established in the envelope thermal properties (Tables 1.1, 1.2 and 1.3), mainly based on stationary thermal properties (the thermal transmittance or the thermal resistance). Regarding Spain and Portugal, the climate classification is obtained by combining the climate severity of each zone in winter and summer [43, 44], limiting the thermal properties according to the winter zone. On the other hand, France presents a slightly different development when making a general climate classification of the whole territory and when establishing limit values of the thermal resistance or of the thermal transmittance of the envelope elements [45, 46]. Likewise, the limit values of the thermal properties of these regulations are constantly changing due to the continuing review, and the values are more and more effective. However, these regulations could generate energy inequalities, both within a country and between neighbouring countries. In this regard, establishing limitations according to the climate zone, although it is an easy task from the

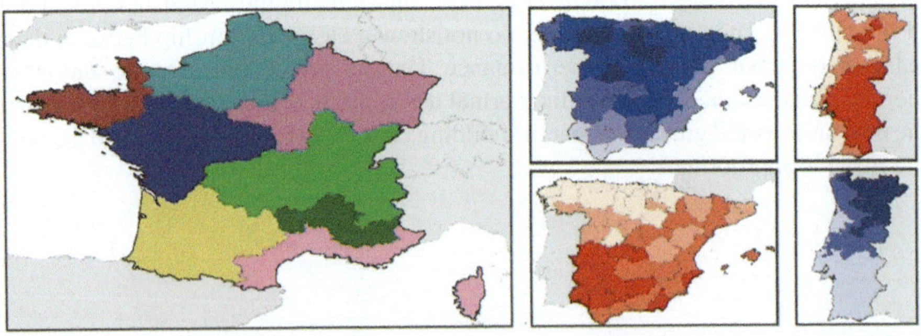

Fig. 1.6 Climate zones of France, Spain, and Portugal

Table 1.1 Maximum thermal transmittance values of the opaque and glazed elements of the building envelope established by the Spanish regulation

Element	Maximum thermal transmittance [W/(m^{2}K)]				
	Winter climate zone				
	A	B	C	D	E
Wall	1.25	1.00	0.75	0.60	0.55
Elements in contact with the ground	1.25	1.00	0.75	0.60	0.55
Roof	0.80	0.65	0.50	0.40	0.35
Floor in contact with air	0.80	0.65	0.50	0.40	0.35
Window	5.70	4.20	3.10	2.70	2.50

Table 1.2 Maximum thermal transmittance values of the opaque and glazed elements of the building envelope established by the Portuguese regulation

Element	Maximum thermal transmittance [W/(m^{2}K)]					
	Continental Portugal			Autonomous regions		
	Climate zone			Climate zone		
	I1	I2	I3	I1	I2	I3
Walls	0.50	0.40	0.35	0.70	0.60	0.45
Roofs	0.40	0.35	0.30	0.45	0.40	0.35
Windows	2.80	2.40	2.20	2.80	2.40	2.20

Table 1.3 Maximum thermal transmittance values of the opaque and glazed elements of the building envelope established by the French regulation

Element	Minimum thermal resistance [(m^{2}K)/W)]		
	Thermal zone		
	H1a, H1b, and H1c	H2a, H2b, H2c, H2d, and H3 at an altitude greater than 800 m	H3 at an altitude lower than 800 m
Walls	2.9	2.9	2.2
Horizontal roofs	3.3	3.3	3.3
Inclined roofs	4.4	4.3	4.0
Floors	2.7	2.7	2.1

regulatory point of view, could imply that buildings located in the same zone have different energy performance. Climate zones of various countries could be similar to their climate characteristics, although different values are established as regards regulations [41]. Different regulation values are established among similar climate zones of Spain and Portugal, thus generating significant differences in the energy demand. In this regard, the winter climate zone A in Spain and I1 in Portugal is similar in some regions, but the thermal transmittance values proposed by each regulation are different: a value of 1.00 W/m^2K is established for walls in Spain and 0.50 W/m2K in Portugal.

However, none of these countries establishes limitations in the periodic thermal properties. As presented in Sect. 1.3, the stationary thermal transmittance should be analysed together with the periodic thermal properties of the envelope to control the energy demand of a building. However, very few countries consider these thermal properties, including Italy. The Italian legislation has established limitations for the thermal mass and the periodic thermal transmittance of the envelope elements through the Decreto Interministeriale 26 giugno 2015 [47]. This regulation establishes that the limit value for the thermal mass should be greater than 230 kg/m^2, and the periodic thermal transmittance should be lower than 0.12 $W/(m^2K)$. Moreover, the Decreto Ministeriale 26/6/2009 [48] establishes a quality classification of envelopes according to the time shift and the decrement factor. The development of the periodic thermal properties has implied that some authors consider the need for establishing limitations of the periodic thermal properties in the regulation of each country, such as Spain [38]. Despite the possible advantages of controlling these variables in the legislation of the southern European countries (due to the limitations to achieve nZEB in the climates of these countries [18]), today these variables should be voluntarily controlled by architects and engineers (except Italy).

References

1. W. McDonough, M. Braungart, Cradle to cradle. Chem. Eng. News. **80**, 208 (2002). https://doi.org/10.1021/es0326322
2. WWF, Living planet report 2014: Species and spaces, people and places. Res. Accunting. **1**, 180 (2014). https://doi.org/10.1007/s13398-014-0173-7.2
3. Intergovernmental Panel on Climate Change, Climate Change 2007: The Physical Science Basis. Contribution of Working Group I to the Fourth Assessment Report of the Intergovernmental Panel on Climate Change (2007)
4. World Meteorological Organization, United In Science (2019)
5. European Commission, *A Roadmap for Moving to a Competitive Low Carbon Economy in 2050* (Belgium, Brussels, 2011)
6. I. Richardson, M. Thomson, D. Infield, A high-resolution domestic building occupancy model for energy demand simulations. Energy. Build. **40**, 1560–1566 (2008). https://doi.org/10.1016/j.enbuild.2008.02.006
7. I. Richardson, M. Thomson, D. Infield, C. Clifford, Domestic electricity use: A high-resolution energy demand model. Energy. Build. **42**, 1878–1887 (2010). https://doi.org/10.1016/j.enbuild.2010.05.023
8. L. Pérez-Lombard, J. Ortiz, C. Pout, A review on buildings energy consumption information. Energy. Build. **40**, 394–398 (2008). https://doi.org/10.1016/j.enbuild.2007.03.007
9. United Nations Environment Programme, *Building Design and Construction: Forging Resource Efficiency and Sustainable Development*. Sustain Build Clim Initiat (2012)
10. International Energy Agency, World Energy Outlook 2013 (2013)
11. N.A. Kurekci, Determination of optimum insulation thickness for building walls by using heating and cooling degree-day values of all Turkey's provincial centers. Energy. Build. **118**, 197–213 (2016). https://doi.org/10.1016/j.enbuild.2016.03.004

12. E.L. Vine, E. Kazakevicius, Residential energy use in Lithuania: The prospects for energy efficiency. Energy. **24**, 591–603 (1999). https://doi.org/10.1016/S0360-5442(99)00013-4
13. G. Besagni, M. Borgarello, The determinants of residential energy expenditure in Italy. Energy. **165**, 369–386 (2018). https://doi.org/10.1016/j.energy.2018.09.108
14. D. Brounen, N. Kok, J.M. Quigley, Residential energy use and conservation: Economics and demographics. Eur. Econ. Rev. **56**, 931–945 (2012). https://doi.org/10.1016/j.euroecorev.2012.02.007
15. M. Filippini, S. Pachauri, Elasticities of electricity demand in urban Indian households. Energy. Policy. **32**, 429–436 (2004). https://doi.org/10.1016/S0301-4215(02)00314-2
16. European Environment Agency, *Final Energy Consumption by Sector and Fuel (2016)* (Denmark, Copenhagen, 2018)
17. I. Zacà, D. D'Agostino, P.M. Congedo, C. Baglivo, Assessment of cost-optimality and technical solutions in high performance multi-residential buildings in the Mediterranean area. Energy. Build. **102**, 250–265 (2015). https://doi.org/10.1016/j.enbuild.2015.04.038
18. S. Attia, P. Eleftheriou, F. Xeni et al., Overview and future challenges of nearly zero energy buildings (nZEB) design in Southern Europe. Energy. Build. **155**, 439–458 (2017). https://doi.org/10.1016/j.enbuild.2017.09.043
19. G. Costa, Á Sicilia, X Oregi et al., A catalogue of energy conservation measures (ECM) and a tool for their application in energy simulation models. J. Build. Eng. **29**. (2020) https://doi.org/10.1016/j.jobe.2019.101102
20. O. Escorcia, R. García, M. Trebilcock et al., Envelope improvements for energy efficiency of homes in the south-central Chile. Inf. La Construcción. **64**, 563–574 (2012). https://doi.org/10.3989/ic.11.143
21. C. Friedman, N. Becker, E. Erell, Energy retrofit of residential building envelopes in Israel: A cost-benefit analysis. Energy. **77**, 183–193 (2014). https://doi.org/10.1016/j.energy.2014.06.019
22. R. Pacheco, J. Ordóñez, G. Martínez, Energy efficient design of building: A review. Renew. Sustain. Energy. Rev. **16**, 3559–3573 (2012). https://doi.org/10.1016/j.rser.2012.03.045
23. V.R. De Lieto, C. Guattari, L. Evangelisti et al., Building energy performance analysis: A case study. Energy. Build. **87**, 87–94 (2015). https://doi.org/10.1016/j.enbuild.2014.10.080
24. U.T. Aksoy, M. Inalli, Impacts of some building passive design parameters on heating demand for a cold region. Build. Environ. **41**, 1742–1754 (2006). https://doi.org/10.1016/j.buildenv.2005.07.011
25. B. Güçyeter, H.M. Günaydin, Optimization of an envelope retrofit strategy for an existing office building. Energy. Build. **55**, 647–659 (2012). https://doi.org/10.1016/j.enbuild.2012.09.031
26. A. Invidiata, M. Lavagna, E. Ghisi, Selecting design strategies using multi-criteria decision making to improve the sustainability of buildings. Build. Environ. **139**, 58–68 (2018). https://doi.org/10.1016/j.buildenv.2018.04.041
27. D. Qian, Y. Li, F. Niu, Z. O'Neill, Nationwide savings analysis of energy conservation measures in buildings. Energy. Convers. Manag. **188**, 1–18 (2019). https://doi.org/10.1016/j.enconman.2019.03.035
28. N. Aste, F. Leonforte, M. Manfren, M. Mazzon, Thermal inertia and energy efficiency—Parametric simulation assessment on a calibrated case study. Appl. Energy. **145**, 111–123 (2015)
29. Z. Yilmaz, Evaluation of energy efficient design strategies for different climatic zones: Comparison of thermal performance of buildings in temperate-humid and hot-dry climate. Energy. Build. **39**, 306–316 (2007). https://doi.org/10.1016/j.enbuild.2006.08.004
30. E. Rodrigues, M.S. Fernandes, A.R. Gaspar et al., Thermal transmittance effect on energy consumption of Mediterranean buildings with different thermal mass. Appl. Energy. **252**, 113437 (2019)
31. A. Dodoo, L. Gustavsson, R. Sathre, Effect of thermal mass on life cycle primary energy balances of a concrete-and a wood-frame building. Appl. Energy. **92**, 462–472 (2012)
32. International Organization for Standardization, ISO 13786:2017 *Thermal Performance of Building Components—Dynamic Thermal Characteristics—Calculation Methods* (2017)

33. N. Aste, A. Angelotti, M. Buzzetti, The influence of the external walls thermal inertia on the energy performance of well insulated buildings. Energy. Build. **41**, 1181–1187 (2009). https://doi.org/10.1016/j.enbuild.2009.06.005
34. C. Di Perna, F. Stazi, A.U. Casalena, M. D'Orazio, Influence of the internal inertia of the building envelope on summertime comfort in buildings with high internal heat loads. Energy. Build. **43**, 200–206 (2011). https://doi.org/10.1016/j.enbuild.2010.09.007
35. E. Kossecka, J. Kosny, Influence of insulation configuration on heating and cooling loads in a continuously used building. Energy. Build. **34**, 321–331 (2002). https://doi.org/10.1016/S0378-7788(01)00121-9
36. F. Stazi, C. Di Perna, P. Munafò, Durability of 20-year-old external insulation and assessment of various types of retrofitting to meet new energy regulations. Energy. Build. **41**, 721–731 (2009). https://doi.org/10.1016/j.enbuild.2009.02.008
37. F. Stazi, G. Ulpiani, M. Pergolini, C. Di Perna, The role of areal heat capacity and decrement factor in case of hyper insulated buildings: An experimental study. Energy. Build. **176**, 310–324 (2018)
38. D. Bienvenido-Huertas, C. Rubio-Bellido, JA Pulido-Arcas, A Pérez-Fargallo (2020) Towards the implementation of periodic thermal transmittance in Spanish building energy regulation. J. Build. Eng. **31**. https://doi.org/10.1016/j.jobe.2020.101402
39. International Organization for Standardization, *ISO 6946:2007—Building Components and Building Elements—Thermal Resistance and Thermal Transmittance—Calculation Method* (Switzerland, Geneva, 2007)
40. M.P. del Pablo-Romero, R. Pozo-Barajas, R. Yñiguez, Global changes in residential energy consumption. Energy Policy **101**, 342–352 (2017). https://doi.org/10.1016/j.enpol.2016.10.032
41. D. Bienvenido-Huertas, M. Oliveira, C. Rubio-Bellido, D. Marín, A comparative analysis of the international regulation of thermal properties in building envelope. Sustainability. **11**, 5574 (2019). https://doi.org/10.3390/su11205574
42. B. Rodríguez-Soria, J. Domínguez-Hernández, J.M. Pérez-Bella, J.J. Del Coz-Díaz, Review of international regulations governing the thermal insulation requirements of residential buildings and the harmonization of envelope energy loss. Renew. Sustain. Energy. Rev.s **34**, 78–90 (2014). https://doi.org/10.1016/j.rser.2014.03.009
43. Gobierno de España, Real Decreto 314/2006, de 17 de marzo, por el que se aprueba el Código Técnico de la Edificación (2006)
44. Ministério das Obras Públicas Transportes e Comunicações, Decreto-Lei n.o 80/2006. 2468–2513 (2006)
45. Republique Française, Code de la construction et de l'habitation (2019)
46. Republique Française, Arrêté du 22 mars 2017 modifiant l'arrêté du 3 mai 2007 relatif aux caractéristiques thermiques et à la performance énergétique des bâtiments existants (2017)
47. Ministri dell'ambiente e della tutela del territorio e del mare delle infrastrutture e dei trasporti e per la semplificazione e la pubblica amministrazione, Decreto interministeriale 26 giugno 2015—Adeguamento linee guida nazionali per la certificazione energetica degli edifici (2015)
48. Ministero dello Sviluppo Economico, Decreto Ministeriale, 26/6/2009—Ministero dello Sviluppo Economico Linee guida nazionali per la certificazione energetica degli edifici (2009)

Chapter 2
Methods to Assess the Thermal Properties of the Building Envelope

2.1 Introduction

The thermal properties of the building envelope are crucial in building energy performance. The variation of envelope thermal values (e.g. through the regulation of a country) directly influences building energy performance [1].

The existing building stock is characterized by having envelopes with poor thermal performance [2]. Moreover, this becomes more important in walls as walls are both the element with the greatest surface in contact with the external air [3] and where the greatest energy losses take place. The thermal performance of walls is deficient because of their design and the ageing of materials [4]. The improvement of the wall thermal performance would reduce building heat losses between 10 and 45% [5]. Consequently, the energy consumption would also be reduced.

Thus, designing ECMs to reduce the building energy consumption is essential. However, it is difficult to know the energy performance of existing buildings. To reduce the uncertainty of the characteristics of an envelope, a correct thermal characterization of the envelope is essential. Determining correctly the thermal properties implies not overestimating the energy consumption. In addition, a mistaken thermal transmittance value of a wall influences the calculation of other aspects related to the energy rehabilitation (e.g. the range of thermal comfort hours), thus implying the proposal of measures not adapted to reality and the increase of payback periods.

Architects, engineers, and other professionals responsible for the works based on the energy improvement of the existing building stock should correctly characterize the thermal properties to reduce energy consumption and CO_2 emissions in accordance with the new goals and priorities of the twenty-first century society. As seen in Chap. 1, the stationary thermal transmittance and the periodic thermal properties are those controlling the building energy demand appropriately. These properties could be determined by procedures such as the theoretical calculation methods (to assess the stationary thermal transmittance and the periodic thermal properties) or in-situ tests (to assess only the stationary thermal transmittance) [6].

D. Bienvenido-Huertas and C. Rubio-Bellido, *Optimization of the Characterization of the Thermal Properties of the Building Envelope*, SpringerBriefs in Applied Sciences and Technology, https://doi.org/10.1007/978-3-030-63629-6_2

2.2 Methods to Assess Static Thermal Properties

As seen in Chap. 1, the stationary thermal transmittance is the most important variable among the stationary thermal properties to control building energy performance. This variable could be determined by the theoretical procedure included in ISO 6946 or by tests.

2.2.1 *Theoretical Methods*

The ISO 6946 method [7] is the theoretical estimation procedure to calculate the stationary thermal transmittance of a wall through the characteristics of its layers (Fig. 2.1). Particularly, this method is based on determining the thermal resistance of each layer (obtained through the thermal conductivity and thickness) and the internal and external surface thermal resistances of walls using the following equation:

$$U = \frac{1}{R_{s,\mathrm{in}} + \sum_{i=1}^{n} \frac{s_i}{\lambda_i} + R_{s,\mathrm{ext}}} \tag{2.1}$$

where λ_i and s_i are the thermal conductivity and thickness of each layer of the wall, respectively, and $R_{s,\mathrm{in}}$ and $R_{s,\mathrm{ext}}$ are the internal and external surface thermal resistances, respectively. $R_{s,\mathrm{in}}$ and $R_{s,\mathrm{ext}}$ are obtained through the tabulated values in ISO 6946 that depend on accurate boundary conditions in terms of heat transfer by convection and radiation. As for walls, the values are 0.13 m^2·K/W for $R_{s,\mathrm{in}}$ and 0.04 m^2·K/W for $R_{s,\mathrm{ext}}$.

The main advantages of the method are that it is a simple calculation procedure and tests are not required to determine the stationary thermal transmittance, so it

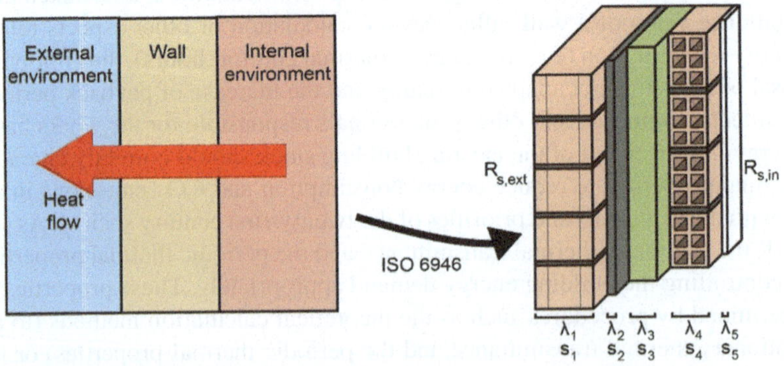

Fig. 2.1 Scheme of the theoretical foundation of ISO 6946

is considerably used in the design phase. In addition, it is included in the energy regulation of several countries as the methodology to determine the stationary thermal transmittance [8]. However, its use in existing buildings is not advisable because, in most cases, the number, type, and thickness of the wall layers are unknown [9]. Thus, the process to determine the layers is something of a challenge and only the use of reliable technical documentation [10] or the endoscopy [9, 11, 12] would obtain accurate information on the design of envelopes. However, the difficulties to access to the technical documentation of a project (particularly in old buildings) or the damages generated by the endoscopic techniques make difficult to characterize the stationary thermal transmittance through ISO 6946 correctly.

In addition, the method has several limitations due to the thermal conductivity values of the layers. One of the most significant limitations is the use of tabulated values, such as those included in databases. In these cases, the thermal conductivity could oscillate between a minimum and a maximum value according to a property (e.g. density). The use of one or another limit value could significantly vary the result of the stationary thermal transmittance. For this reason, Ficco et al. [9] proposed to average the maximum ($U_{i,max}$) and minimum values ($U_{i,min}$) related to the wall according to the distribution of the thermal conductivity values for each layer:

$$U = \frac{U_{i,max} + U_{i,min}}{2} \tag{2.2}$$

Another limitation related to the thermal conductivity is the variation presented by the materials due to several factors. Aspects such as the presence of humidity, the ageing of materials or the environmental conditions could vary the thermal conductivity in comparison to the tabulated values [6]. Regarding the variations due to environmental conditions, one of the reasons causing this limitation is that the thermal conductivity values vary according to humidity and the environmental temperature [13–16, 17]. Most databases, such as the Constructive Elements Catalogue in Spain (linked to the Spanish Technical Building Code), use fixed environmental values to calculate the thermal properties of materials according to ISO 10456 [18]. However, the variations of the thermal conductivity values in various climate regions are not considered. Consequently, Pérez-Bella et al. [19, 20, 21] established a simplified procedure to apply ISO 10456 by using some conductivity correction factors (CCFs) obtained for each province capital in Spain according to the environmental conditions (Fig. 2.2). These correction factors simplify the application of ISO 10456 by combining a humidity conversion factor ($F_{H\text{correction}}$) and a temperature conversion factor ($F_{T\text{correction}}$) (Eq. (2.3)), which are applied to the thermal conductivity of the material to determine the actual value (Eq. (2.4)).

$$CCF = F_{H\text{correction}} \cdot F_{T\text{correction}} \tag{2.3}$$

$$\lambda_{CCF} = \lambda \cdot CCF \tag{2.4}$$

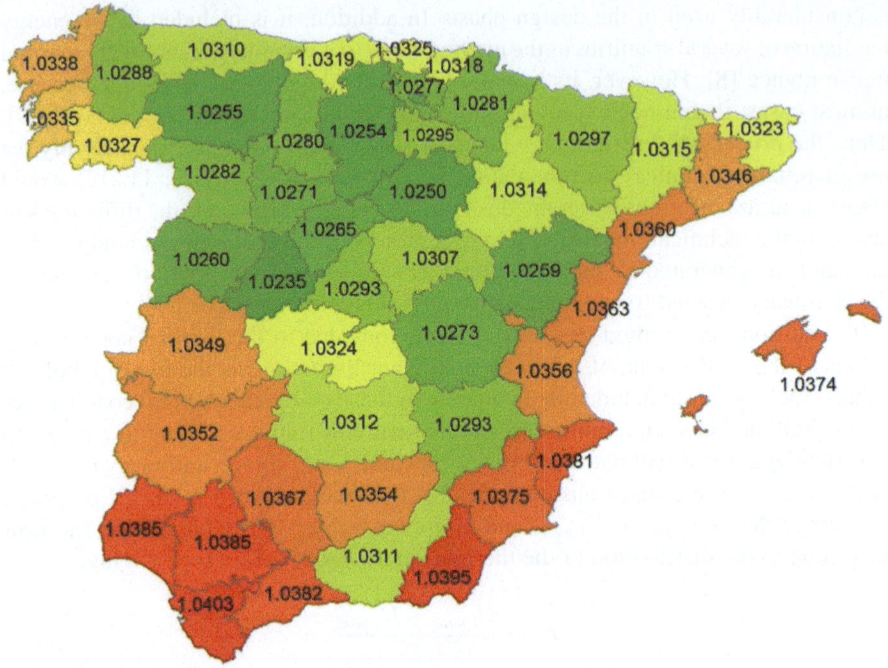

Fig. 2.2 Map with the CCF values for the theoretical calculation

where λ_{CCF} is the thermal conductivity of the material by applying the correction of the CCF.

Despite these limitations, this method is widely used in the design phase as a justification of the fulfilment in many countries of the state regulation on energy efficiency [8]. It is also used to validate the results of the experimental methods thanks to one of the criteria included in ISO 9869-1: when the percentage deviation obtained with Eq. (2.5) is lower than 20%, the result is representative.

$$\sigma = \frac{U_{\text{in situ}} - U_{6946}}{U_{6946}} \tag{2.5}$$

where U_{6946} is the thermal transmittance obtained through ISO 6946, and $U_{\text{in situ}}$ is the thermal transmittance obtained experimentally.

Although this method is very significant, its use in existing buildings is something of a challenge. Aspects such as the unknowing of the design of the wall [22–25], the ageing of materials [17, 26] or humidity [27] could obtain percentage deviations greater than 20%. Thus, to guarantee a most appropriate use of the method in existing buildings, new methodologies should be available to estimate accurately such stationary thermal transmittance value, thus removing the main barriers to use the method.

2.2.2 *Experimental Methods*

The experimental methods to characterize the stationary thermal transmittance could be divided into three groups [28]: the heat flow meter method [29], the thermometric method [30, 31, 32] and the quantitative method through infrared thermography [33].

2.2.2.1 The Heat Flow Meter Method

The heat flow meter method (HFM) is to date the only standardized method through ISO 9869-1 [29]. This method consists in obtaining the thermal transmittance value by measuring the heat flux (q_j) that goes through the wall and the interior ($T_{\text{in},j}$) and exterior ($T_{\text{ext},j}$) ambient temperatures that divide the wall:

$$U = \frac{\sum_{j=1}^{n} q_j}{\sum_{j=1}^{n} \left(T_{\text{in},j} - T_{\text{ext},j}\right)} \tag{2.6}$$

For this purpose, measurements are carried out in the wall lasting from 72 h [9] to over 1 week [34–36]. Measurements are mainly carried out with a heat flux plate and temperature sensors. Likewise, the use of an infrared camera would guarantee that probes are placed far from the zones affected by thermal heterogeneities [9, 22, 37, 38]. Probes are placed by locating the heat flux plate 150 cm above the floor [39] and far from those zones with thermal discontinuities [3, 25, 40]. Air temperature probes should be placed far from radiation sources to avoid distortions in measurements, always guaranteeing 30 cm from the wall to avoid convective effects [41]. Figure 2.3 presents the criteria for placing the probes.

HFM is standardized through ISO 9869-1, so it is more and more used by professionals and researchers. Regarding this last aspect, many studies have analysed the possibilities of using this method in existing buildings. Some examples are Lucchi [17, 26] and Rotilio et al. [42]. However, most studies are focused on addressing the limitations of HFM. These limitations could be related to metrological and operational problems, the influence of environmental factors, and the difficulties to analyse the results. Regarding the metrological and operational problems, Cesaratto et al. [43], Desogus et al. [11] and Trethowen [44] reported that the main existing deviations in the results are due to the disturbance of the heat flux by placing the probes in the wall. Likewise, the errors by using the plate could be significant [45]. Cucumo et al. [46] and Meng et al. [47] showed that the error to determine the stationary thermal transmittance by placing the plate could oscillate between 26 and 30%.

As for the environmental factors, a high temperature gradient is essential to obtain representative results. Desogus et al. [11] obtained an uncertainty of 10% for a temperature difference of 10 °C between the interior and the exterior, obtaining lower percentage errors when the thermal gradient was greater than 10 °C. This

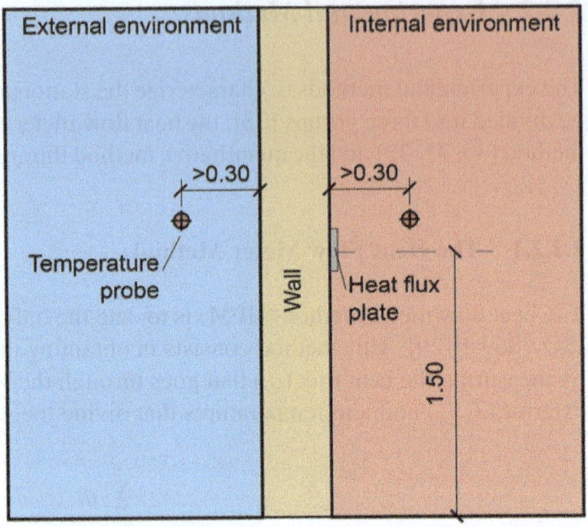

Fig. 2.3 Criteria for placing the probes with HFM

requirement was confirmed by Refs. [35, 9, 48, 49]. In addition, this requirement should be extended the whole test time, so there could be limitations in the warmest climate zones [50]. Another essential aspect to guarantee that HFM is correctly carried out is the presence of water in the wall as precipitation, snow, and humidity significantly influence the heat flux of the wall, thus varying the results up to 71% [27, 51, 52]. The orientation of the wall should also be considered. Ahmad et al. [34] showed that walls not facing north present a greater heat flux that those facing north, obtaining deviations greater than 37.3%. Moreover, the use of heating systems and their operation cycle when performing tests is an aspect to be considered. A common practice to guarantee a high thermal gradient is the use of heating systems. The operation cycles of the heating system vary the surface temperature that modify the heat flux through the wall, with significant variations in the convective heat exchange [53]. Likewise, placing the probes far from these systems is crucial. Guattari et al. [54] recommended to place the probes at least at 130 cm from heating systems.

Finally, the data analysis method could significantly vary the representation of results. Cesaratto and Carli [55] established that the representation of the results could be improved with a deliberate data filtering in those instances with a high thermal gradient. With the filtering method, the results obtained with the same dataset could vary up to 20% [55]. Likewise, the formulation could also vary the thermal transmittance results obtained with HFM, such as the case of the dynamic methods [56] or the alternative formulation included in ISO 9869-1 to carry out heat storage corrections:

$$U = \frac{\sum_{j=1}^{n} q_j - \frac{(F_{\text{in}} \cdot \delta T_{\text{in}} + F_{\text{ext}} \cdot \delta T_{\text{ext}})}{\Delta t}}{\sum_{j=1}^{n} \left(T_{\text{in},j} - T_{\text{ext},j}\right)} \tag{2.7}$$

where F_{in} is the total internal thermal mass factor, δT_{in} is the difference between the average internal air temperature from the 24 h before the measurement and the average internal air temperature from the first 24 h of the test, F_{ext} is the total external thermal mass factor, δT_{ext} is the difference between the average external air temperature from the 24 h before the measurement and the average external air temperature from the first 24 h of the test, and Δt is the time interval between the measurements.

2.2.2.2 The Thermometric Method

The thermometric method (THM) is another non-destructive method recently developed [28]. This method is based on the Newton's law of cooling, which establishes that when the temperature difference between a wall and the environment is not high, the heat rate transferred by conduction, convection, and radiation is virtually proportional to the temperature difference between the wall and the environment:

$$\frac{dQ}{dt} = S \cdot h \cdot [T - T_{\text{environment}}] \tag{2.8}$$

where Q is the thermal energy, S is the heat transfer area of the wall,h is the heat transfer coefficient, T is the temperature of the wall, and $T_{\text{environment}}$ is the environmental temperature. With Eq. (2.8), the heat flux could be modified by a new expression (Eq. (2.9)), and with Eq. (2.9), the equation of HFM (Eq. (2.6)) could be modified, thus obtaining the expression used for THM (Eq. (2.10).

$$q_j = h_{\text{in}} \cdot \left(T_{\text{in},j} - T_{s,\text{in},j}\right) \tag{2.9}$$

$$U = \frac{\sum_{j=1}^{n} h_{\text{in}} \left(T_{\text{in},j} - T_{s,\text{in},j}\right)}{\sum_{j=1}^{n} \left(T_{\text{in},j} - T_{\text{ext},j}\right)} \tag{2.10}$$

where h_{in} is the total heat transfer coefficient.

Therefore, the main difference of THM in comparison with HFM is that measuring the heat flux of the wall is not required (Fig. 2.4). To perform the test, the heat flux place is replaced by surface temperature probes. The criteria for placing these probes

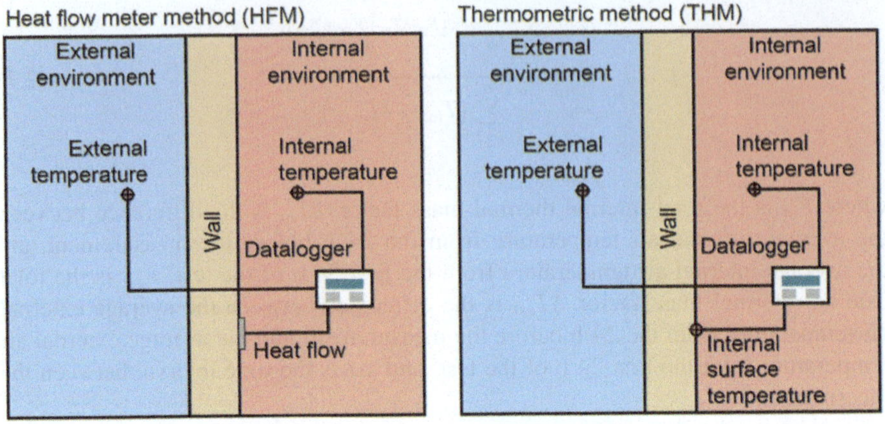

Fig. 2.4 Sketch of the difference between HFM and THM

follow the same height criterion as in HFM with respect to the floor, although a maximum distance of 15 cm among the probes and a distance of 2 cm from the mortar joints of the brick layer should be guaranteed [47]. Likewise, the criteria for the test duration are the same as those for HFM, oscillating between 72 and 168 h [31, 6], although if steady conditions are guaranteed, monitorings could be performed only in the small hours, with an average deviation of 2.63% in comparison with long measurements [57].

This method is widely used by professionals, and even by accredited laboratories in some countries [31]. Its main advantage is that there are not measurement errors due to the use of the heat flow meter probe described for HFM [6]. In this regard, the error presented in the results by the surface temperature sensors is 20% lower than the error presented by the heat flux plate [47].

However, some operational and environmental requirements of HFM are also applied to THM, including the need for a high thermal gradient, the lack of radiant elements, thermal bridges, the orientation of the wall or the presence of humidity [31, 57, 32, 28]. Despite of this, the main limitation of the method is the value used for h_{in}. THM use a value of 7.69 W/(m^2·K), which is obtained from the inverse of the internal surface thermal resistance ($R_{s,in}$) for horizontal flows of ISO 6946. As Annex A of this standard indicates, this value represents walls whose materials have an emissivity close to 0.9 and indoor boundary conditions oscillate between 20 and 25 °C. As a conventional wall generally fulfils both conditions, this value could be applied to most walls of the building stock. However, if these boundary conditions are not met, there could be deviations in the results of the method.

2.2.2.3 Quantitative Method Through Infrared Thermography

Another non-destructive method recently developed is the quantitative method through infrared thermography. Traditionally, the infrared thermography has been widely used, both from the medical and military sides. In the field of energy audit, the infrared thermography has been mainly used to analyse building envelopes qualitatively [58] in various ways:

- The detection of thermal anomalies [27, 59].
- The detection of thermal bridges [60, 61].
- Air infiltrations together with the blower door test [62, 63].

However, recent studies have analysed its application to determine the thermal transmittance while a qualitative analysis is carried out. These studies are based on the possibility of using the infrared thermography to measure some of the variables used in the formulation approaches, such as the surface temperature ($T_{s,\mathrm{in}}$), the reflected apparent temperature (T_{refl}) or the emissivity (ε).

The approaches of the quantitative method could be divided into two groups according to the place from where measurements are carried out with the infrared camera: from the exterior [35, 64, 39, 65] and from the interior Madding [66], Fokaides and Kalogirou [67], and Tejedor et al. [68, 69]. Moreover, each researcher used a different formulation approach. These differences are mainly based on the value or equation used for the convective coefficient [70, 71].

The first approach was established by Madding [66], who determined the stationary thermal transmittance by using the thermography from the interior with a formulation based on the Stephan–Boltzmann law (Eq. (2.11)). Fokaides and Kalogirou [67] developed a similar proposal. In this case, Eq. (2.11) was modified by only using the third power of $T_{s,\mathrm{in}}$ (Eq. (2.12). As for h_{in}, different approaches were used: Madding used the correlations of Earle [72] and Holman [73], and Fokaides and Kalogirou used the tabulated value from ISO 6946. In addition, the recent studies by Tejedor et al. [68, 69] proposed a new formulation approach based on adimensional numbers (Eq. (2.13)).

$$U = \frac{4\varepsilon\sigma\left(\frac{T_{s,\mathrm{in}}+T_{\mathrm{refl}}}{2}\right)^3\left(T_{s,\mathrm{in}} - T_{\mathrm{refl}}\right) + h_{\mathrm{in}} \cdot \left(T_{s,\mathrm{in}} - T_{\mathrm{in}}\right)}{T_{\mathrm{in}} - T_{\mathrm{ext}}} \tag{2.11}$$

$$U = \frac{4 \cdot \varepsilon \cdot \sigma \cdot T_{s,\mathrm{in}}^3 \cdot \left(T_{s,\mathrm{in}} - T_{\mathrm{refl}}\right) + h_{\mathrm{in}} \cdot \left(T_{s,\mathrm{in}} - T_{\mathrm{in}}\right)}{T_{\mathrm{in}} - T_{\mathrm{ext}}} \tag{2.12}$$

$$U = \frac{\varepsilon \cdot \sigma \cdot \left(T_{\mathrm{refl}}^4 - T_{s,\mathrm{in}}^4\right) + \frac{k \cdot \left\{0.825+0.325 \cdot Ra^{1/6}\right\}^2}{L} \cdot \left(T_{\mathrm{in}} - T_{s,\mathrm{in}}\right)}{T_{\mathrm{in}} - T_{\mathrm{ext}}} \tag{2.13}$$

where σ is the Stefan–Boltzmann constant, Ra_L is the Rayleigh number, k is the thermal conductivity of the air, and L is the height of the wall.

In addition, the recent studies by Albatici et al. [35, 64, 39, 65] developed approaches of the quantitative method from the exterior. Dall'O et al. [65] used the convection correlation published by Watanabe without considering the radiation coefficient (Eq. (2.14)), and Albatici et al. [35, 64, 39] applied a formulation used by the simplified correlation of Watanabe considering the radiation coefficient (Eq. (2.15)). In both approaches, the measurement of the local wind speed was crucial (v).

$$U = \frac{(5.8 + 3.8054 \cdot v) \cdot (T_{s,\text{ext}} - T_{\text{ext}})}{T_{\text{in}} - T_{\text{ext}}} \tag{2.14}$$

$$U = \frac{\varepsilon \cdot \sigma \cdot (T_{s,\text{ext}}^4 - T_{\text{ext}}^4) + 3.8054 \cdot v \cdot (T_{s,\text{ext}} - T_{\text{ext}})}{T_{\text{in}} - T_{\text{ext}}} \tag{2.15}$$

There are many approaches of the method, thus implying a limitation since results could vary [70, 71]. However, performing tests quickly is an important advantage in comparison with HFM and THM. The following criteria are required to perform tests (Fig. 2.5):

- The infrared camera should be placed at 150 cm from the wall and not perpendicular to it [69, 74].
- All the elements required for the measurement, such as the reflector, should be placed 150 cm above the floor [69].
- The hot-wire anemometer should be placed at 10 cm from the wall surface [39].

Regarding other limitations of this method, there are similarities with HFM and THM [6]. One of the main limitations are the environmental requirements, which are characterized by the need for a high thermal gradient [33]. Moreover, a low

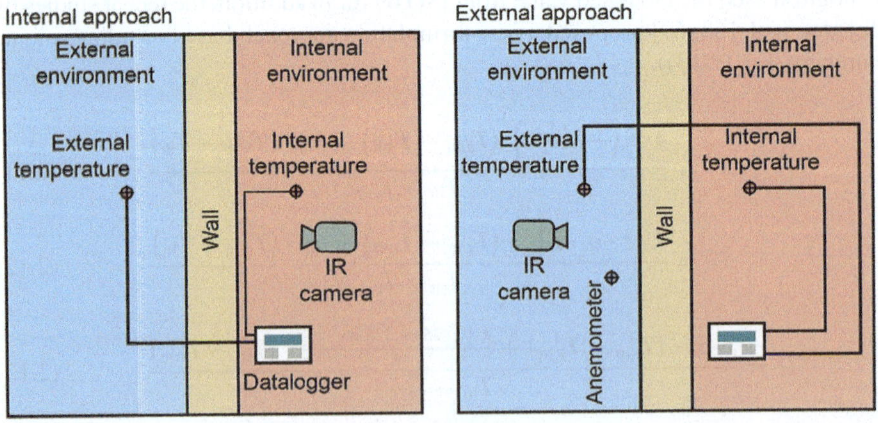

Fig. 2.5 Criteria for placing the equipment and probes of the quantitative method through infrared thermography

differential between the reflected apparent temperature and the external temperature obtains more adjusted results. Thus, the ideal test conditions are in winter because achieving a high thermal gradient in summer is something of a challenge [67, 69]. Other climatic parameters such as wind [75], solar radiation [76] or rainfalls [69] significantly influence tests. In addition, the typology of the element analysed (simple or with several layers) affects the limitations of the results obtained through the different methods [68].

2.3 Methods to Assess Periodic Thermal Properties

The periodic thermal properties of walls are characterized through ISO 13786 [77]. This standard is based on the study by Carslaw and Jaeger [78], which analysed the sinusoidal relationship between the heat flux and the internal and external temperatures. Thus, ISO 13786 considers a sinusoidal variation throughout the time of the external air temperature of envelopes, which generates sinusoidal variations in both the heat flux and the internal temperature (Fig. 2.6). This is the only appropriate method to characterize periodic thermal properties. There are not reliable experimental methods to characterize these variables, despite some studies such as that conducted by Pernigotto et al. [79] through hot box, although its use in actual walls have difficulties.

ISO 13786 is a theoretical calculation procedure with the same limitations as ISO 6946. The design of the wall and the thermal properties of its layers should be known. However, more material properties should be known than in ISO 6946 as the density (ρ) and the specific thermal capacity (c) are required. The calculation procedure consists of heat transfer matrix operations of each material of the wall (Z_{mn}) (Eq. (2.16)).

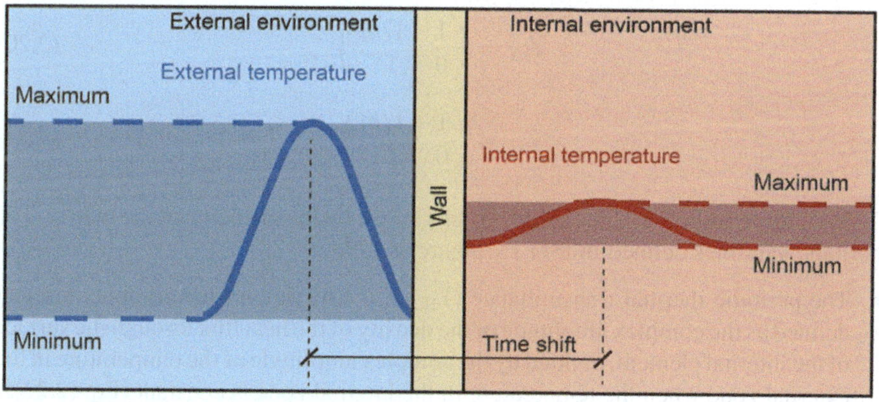

Fig. 2.6 Scheme of the theoretical foundation of ISO 13786

$$Z_{mn} = \begin{pmatrix} Z_{11} & Z_{12} \\ Z_{21} & Z_{22} \end{pmatrix}$$

$$Z_{11} = Z_{22} = \cosh(\xi)\cos(\xi) + j \cdot \mathrm{senh}(\xi)\mathrm{sen}(\xi)$$

$$Z_{12} = -\frac{\delta}{2\lambda}\{\mathrm{senh}(\xi)\cos(\xi) + \cosh(\xi)\mathrm{sen}(\xi)$$
$$+j \cdot [\cosh(\xi)\mathrm{sen}(\xi) - \mathrm{senh}(\xi)\cos(\xi)]\}$$

$$Z_{21} = -\frac{\lambda}{\delta}\{\mathrm{senh}(\xi)\cos(\xi) - \cosh(\xi)\mathrm{sen}(\xi)$$
$$+j \cdot [\cosh(\xi)\mathrm{sen}(\xi) + \mathrm{senh}(\xi)\cos(\xi)]\} \tag{2.16}$$

where δ is the periodic penetration depth of a thermal wave in the material (Eq. (2.17)), and ξ is the relationship between the thickness and δ.

$$\delta = \sqrt{\frac{\lambda T}{\pi \rho c}} \tag{2.17}$$

The total heat transfer matrix (Z) is obtained by multiplying the heat transfer matrices of each material of the wall (Eq. (2.18)) and is used to obtain the heat transfer matrix between two environments (Z_{ee}) (Eq. (2.19)), applying two heat transfer matrices for the boundary conditions, both external (Z_{sN}) (Eq. (2.20)) and internal (Z_{s1}) (Eq. (2.21)).

$$Z = \begin{pmatrix} Z_{11} & Z_{12} \\ Z_{21} & Z_{22} \end{pmatrix} = \prod_{i=N}^{1} Z_i \tag{2.18}$$

$$Z_{ee} = Z_{sN} \cdot Z \cdot Z_{s1} \tag{2.19}$$

$$Z_{sN} = \begin{pmatrix} 1 & -1/h_e \\ 0 & 1 \end{pmatrix} \tag{2.20}$$

$$Z_{s1} = \begin{pmatrix} 1 & -1/h_i \\ 0 & 1 \end{pmatrix} \tag{2.21}$$

Various periodic variables could be determined by the elements of the matrix. The periodic variables defined in ISO 13786 are as follows:

- The periodic thermal transmittance (Y_{12}) is the module of the complex number defined as the complex amplitude of the density of the heat flux through the surface of the internal element, divided by the complex amplitude of the temperature in the external zone when the temperature in the internal zone is constant (Eq. (2.22)).

- Time shift periodic thermal transmittance (φ) is the period between the maximum amplitude of a cause and the maximum amplitude of its effect related to the periodic thermal transmittance (Eq. (2.23)).
- The decrement factor (f) is the quotient between the module of the periodic thermal transmittance and the stationary thermal transmittance (Eq. (2.24)).
- The internal thermal admittance (Y_{11}) is the module of the complex number defined as the complex amplitude of the density of the heat flux through the surface of the component adjacent to the internal zone, divided by the complex amplitude of the temperature in the same zone when the temperature in the internal zone is constant (Eq. (2.25)).
- The time shift internal side (φ_{11}) is the period between the maximum amplitude of a cause and the maximum amplitude of its effect related to the internal thermal admittance (Eq. (2.26)).
- The external thermal admittance (Y_{22}) is the module of the complex number defined as the complex amplitude of the density of the heat flux through the surface of the component adjacent to the external zone, divided by the complex amplitude of the temperature in the same zone when the temperature in the external zone is constant (Eq. (2.27)).
- The time shift external side (φ_{22}) is the period between the maximum amplitude of a cause and the maximum amplitude of its effect related to the external thermal admittance (Eq. (2.28)).
- The internal areal heat capacity (k_1) is the capacity to accumulate heat in the internal side of the element (Eq. (2.29)).
- The external areal heat capacity (k_2) is the capacity to accumulate heat in the external side of the element (Eq. (2.30)).

$$Y_{12} = -\frac{1}{Z_{12}} \tag{2.22}$$

$$\varphi = \frac{T}{2\pi}\arg(Z_{12}) \tag{2.23}$$

$$f = \frac{|Y_{12}|}{U} \tag{2.24}$$

$$Y_{11} = -\frac{Z_{11}}{Z_{12}} \tag{2.25}$$

$$\varphi_{11} = \frac{T}{2\pi}\arg(Y_{11}) \tag{2.26}$$

$$Y_{22} = -\frac{Z_{22}}{Z_{12}} \tag{2.27}$$

$$\varphi_{22} = \frac{T}{2\pi}\arg(Y_{22}) \tag{2.28}$$

$$k_1 = \frac{T}{2\pi}\left|\frac{Z_{11}-1}{Z_{12}}\right| \tag{2.29}$$

$$k_2 = \frac{T}{2\pi}\left|\frac{Z_{22}-1}{Z_{12}}\right| \tag{2.30}$$

References

1. D. Bienvenido-Huertas, M. Oliveira, C. Rubio-Bellido, D. Marín, A Comparative analysis of the international regulation of thermal properties in building envelope. Sustainability **11**, 5574 (2019). https://doi.org/10.3390/su11205574
2. F. Kurtz, M. Monzón, B. López-Mesa, Energy and Acoustics Related Obsolescence of Social Housing of Spain's Post-War in Less Favoured Urban Areas. The Case of Zaragoza. Inf la Construcción 67:m021. (2015). https://doi.org/10.3989/ic.14.062
3. R. Adhikari, E. Lucchi, V. Pracchi, *Experimental Measurements on Thermal Transmittance of the Opaque Vertical Walls in the Historical Buildings*. In PLEA2012 Conference, Opportunities, Limits and Needs Towards an environmentally Responsible Architecture (2012)
4. D.A. Waddicor, E. Fuentes, L. Sisó et al., Climate change and building ageing impact on building energy performance and mitigation measures application: A case study in Turin, northern Italy. Build. Environ. **102**, 13–25 (2016). https://doi.org/10.1016/j.buildenv.2016.03.003
5. R. Walker, S. Pavía, Thermal performance of a selection of insulation materials suitable for historic buildings. Build. Environ. **94**, 155–165 (2015). https://doi.org/10.1016/j.buildenv.2015.07.033
6. D. Bienvenido-Huertas, J. Moyano, D. Marín, R. Fresco-Contreras, Review of in situ methods for assessing the thermal transmittance of walls. Renew Sustain Energy Rev. **102**, 356–371 (2019). https://doi.org/10.1016/j.rser.2018.12.016
7. International Organization for Standardization, *ISO 6946:2007—Building Components and Building Elements—Thermal Resistance and Thermal Transmittance—Calculation Method* (Switzerland, Geneva, 2007)
8. B. Rodríguez-Soria, J. Domínguez-Hernández, J.M. Pérez-Bella, J.J. Del Coz-Díaz, Review of international regulations governing the thermal insulation requirements of residential buildings and the harmonization of envelope energy loss. Renew Sustain Energy Rev. **34**, 78–90 (2014). https://doi.org/10.1016/j.rser.2014.03.009
9. G. Ficco, F. Iannetta, E. Ianniello et al., U-value in situ measurement for energy diagnosis of existing buildings. Energy Build. **104**, 108–121 (2015). https://doi.org/10.1016/j.enbuild.2015.06.071
10. I. Ballarini, S.P. Corgnati, V. Corrado, Use of reference buildings to assess the energy saving potentials of the residential building stock: The experience of TABULA project. Energy Policy **68**, 273–284 (2014). https://doi.org/10.1016/j.enpol.2014.01.027
11. G. Desogus, S. Mura, R. Ricciu, Comparing different approaches to in situ measurement of building components thermal resistance. Energy Build. **43**, 2613–2620 (2011). https://doi.org/10.1016/j.enbuild.2011.05.025
12. V. Echarri, A. Espinosa, C. Rizo, Thermal transmission through existing building enclosures: Destructive monitoring in intermediate layers versus non-destructive monitoring with sensors on surfaces. Sensors **17**, 1–24 (2017). https://doi.org/10.3390/s17122848
13. I. Budaiwi, A. Abdou, The impact of thermal conductivity change of moist fibrous insulation on energy performance of buildings under hot–humid conditions. Energy Build. **60**, 388–399 (2013). https://doi.org/10.1016/j.enbuild.2013.01.035

14. F. Domínguez-Muñoz, B. Anderson, J.M. Cejudo-López, A. Carrillo-Andrés, Uncertainty in the thermal conductivity of insulation materials. Energy Build. **42**, 2159–2168 (2010). https://doi.org/10.1016/j.enbuild.2010.07.006
15. I. Gnip, S. Vėjelis, S. Vaitkus, Thermal conductivity of expanded polystyrene (EPS) at 10 °C and its conversion to temperatures within interval from 0 to 50 °C. Energy Build. **52**, 107–111 (2012). https://doi.org/10.1016/j.enbuild.2012.05.029
16. F. Ochs, W. Heidemann, H. Müller-Steinhagen, Effective thermal conductivity of moistened insulation materials as a function of temperature. Int. J. Heat Mass Transf. **51**, 539–552 (2008). https://doi.org/10.1016/j.ijheatmasstransfer.2007.05.005
17. E. Lucchi, Thermal transmittance of historical brick masonries: A comparison among standard data, analytical calculation procedures, and in situ heat flow metre measurements. Energy Build. **134**, 171–184 (2017). https://doi.org/10.1016/j.enbuild.2016.10.045
18. International Organization for Standardization, *ISO 10456:2007—Building Materials and Products—Hygrothermal Properties—Tabulated Design Values and Procedures for Determining Declared and Design Thermal Values* (Switzerland, Geneva, 2007)
19. J.M. Pérez-Bella, J. Domínguez-Hernández, E. Cano-Suñén et al., Adjusting the design thermal conductivity considered by the Spanish building technical code for wall materials. Dyna. **92**, 1–11 (2017). https://doi.org/10.6036/8005
20. J.M. Pérez-Bella, J. Domínguez-Hernández, E. Cano-Suñén et al., Detailed territorial estimation of design thermal conductivity for wall materials in North-Eastern Spain. Energy Build. **102**, 266–276 (2015). https://doi.org/10.1016/j.enbuild.2015.05.025
21. J.M. Pérez-Bella, J. Domínguez-Hernández, E. Cano-Suñén et al., A correction factor to approximate the design thermal conductivity of building materials. Appl. Spanish Walls. Energy Build. **88**, 153–164 (2015). https://doi.org/10.1016/j.enbuild.2014.12.005
22. P. Baker, U-values and Traditional Buildings: In situ Measurements and their Comparisons to Calculated Values (2011)
23. A. Byrne, G. Byrne, A. Davies, A.J. Robinson, Transient and quasi-steady thermal behaviour of a building envelope due to retrofitted cavity wall and ceiling insulation. Energy Build. **61**, 356–365 (2013). https://doi.org/10.1016/j.enbuild.2013.02.044
24. L. Evangelisti, C. Guattari, P. Gori, V.R. De Lieto, In situ thermal transmittance measurements for investigating differences between wall models and actual building performance. Sustainability **7**, 10388–10398 (2015). https://doi.org/10.3390/su70810388
25. C. Rye, C. Scott, The SPAB Research Report 1. U-Value Report. (2012)
26. E. Lucchi, Thermal transmittance of historical stone masonries: A comparison among standard, calculated and measured data. Energy Build. **151**, 393–405 (2017). https://doi.org/10.1016/j.enbuild.2017.07.002
27. G. Litti, S. Khoshdel, A. Audenaert, J. Braet, Hygrothermal performance evaluation of traditional brick masonry in historic buildings. Energy Build. **105**, 393–411 (2015). https://doi.org/10.1016/j.enbuild.2015.07.049
28. M. Teni, H. Krstić, P. Kosiński, Review and comparison of current experimental approaches for in-situ measurements of building walls thermal transmittance. Energy Build. **203**, 109417 (2019). https://doi.org/10.1016/j.enbuild.2019.109417
29. International Organization for Standardization, ISO 9869-1:2014—Thermal insulation—Building elements—In situ Measurement of Thermal Resistance and Thermal Transmittance. Part 1: Heat Flow Metre Method. (Geneva, Switzerland, 2014)
30. B. Vučićević, V. Turanjanin, V. Bakić et al., Experimental and numerical modelling of thermal performance of a residential building in Belgrade. Therm. Sci. **13**, 245–252 (2009). https://doi.org/10.2298/TSCI0904245V
31. D. Bienvenido-Huertas, R. Rodríguez-Álvaro, J.J. Moyano et al., Determining the U-value of walls using the thermometric method: Potentials and limitations. Energies **11**, 1–17 (2018). https://doi.org/10.3390/en11020360
32. S.-H. Kim, J.-H. Lee, J.-H. Kim et al., The feasibility of improving the accuracy of in situ measurements in the air-surface temperature ratio method. Energies **11**, 1–18 (2018). https://doi.org/10.3390/en11071885

33. I. Nardi, E. Lucchi, T. de Rubeis, D. Ambrosini, Quantification of heat energy losses through the building envelope: A state-of-the-art analysis with critical and comprehensive review on infrared thermography. Build. Environ. **146**, 190–205 (2018). https://doi.org/10.1016/j.buildenv.2018.09.050
34. A. Ahmad, M. Maslehuddin, L.M. Al-Hadhrami, In situ measurement of thermal transmittance and thermal resistance of hollow reinforced precast concrete walls. Energy Build. **84**, 132–141 (2014). https://doi.org/10.1016/j.enbuild.2014.07.048
35. R. Albatici, A.M. Tonelli, Infrared thermovision technique for the assessment of thermal transmittance value of opaque building elements on site. Energy Build. **42**, 2177–2183 (2010). https://doi.org/10.1016/j.enbuild.2010.07.010
36. F. Asdrubali, F. D'Alessandro, G. Baldinelli, F. Bianchi, Evaluating in situ thermal transmittance of green buildings masonries: A case study. Case Stud. Constr. Mater. **1**, 53–59 (2014). https://doi.org/10.1016/j.cscm.2014.04.004
37. F. Ascione, N. Bianco, R.F. De Masi et al., Simplified state space representation for evaluating thermal bridges in building: Modelling, application and validation of a methodology. Appl. Therm. Eng. **61**, 344–354 (2013). https://doi.org/10.1016/j.applthermaleng.2013.07.052
38. F. Ascione, N. Bianco, R.F. De Masi et al., Experimental validation of a numerical code by thin film heat flux sensors for the resolution of thermal bridges in dynamic conditions. Appl. Energy **124**, 213–222 (2014). https://doi.org/10.1016/j.apenergy.2014.03.014
39. R. Albatici, A.M. Tonelli, M. Chiogna, A comprehensive experimental approach for the validation of quantitative infrared thermography in the evaluation of building thermal transmittance. Appl. Energy **141**, 218–228 (2015). https://doi.org/10.1016/j.apenergy.2014.12.035
40. L. Zalewski, S. Lassue, D. Rousse, K. Boukhalfa, Experimental and numerical characterization of thermal bridges in prefabricated building walls. Energy Convers Manag. **51**, 2869–2877 (2010). https://doi.org/10.1016/j.enconman.2010.06.026
41. J.M. Andújar Márquez, M.Á. Martínez Bohórquez, S. Gómez Melgar, A new metre for cheap, quick, reliable and simple thermal transmittance (U-Value) measurements in buildings. Sensors **17**, 1–18 (2017). https://doi.org/10.3390/s17092017
42. M. Rotilio, F. Cucchiella, P. De Berardinis, V. Stornelli, Thermal transmittance measurements of the historical masonries: some case studies. Energies **11**, 2987 (2018). https://doi.org/10.3390/en11112987
43. P.G. Cesaratto, M. De Carli, S. Marinetti, Effect of different parameters on the in situ thermal conductance evaluation. Energy Build. **43**, 1792–1801 (2011). https://doi.org/10.1016/j.enbuild.2011.03.021
44. H. Trethowen, Measurement errors with surface-mounted heat flux sensors. Build. Environ. **21**, 41–56 (1986). https://doi.org/10.1016/0360-1323(86)90007-7
45. C. Peng, Z. Wu, In situ measuring and evaluating the thermal resistance of building construction. Energy Build. **40**, 2076–2082 (2008). https://doi.org/10.1016/j.enbuild.2008.05.012
46. M. Cucumo, V. Ferraro, D. Kaliakatsos, M. Mele, On the distortion of thermal flux and of surface temperature induced by heat flux sensors positioned on the inner surface of buildings. Energy Build. **158**, 677–683 (2018). https://doi.org/10.1016/j.enbuild.2017.10.034
47. X. Meng, B. Yan, Y. Gao et al., Factors affecting the in situ measurement accuracy of the wall heat transfer coefficient using the heat flow metre method. Energy Build. **86**, 754–765 (2015). https://doi.org/10.1016/j.enbuild.2014.11.005
48. K. Gaspar, M. Casals, M. Gangolells, Energy and buildings In situ measurement of walls with a low U-value: Avoiding deviations. Energy Build. **170**, 61–73 (2018). https://doi.org/10.1016/j.enbuild.2018.04.012
49. V. Gori, C.A. Elwell, Estimation of thermophysical properties from in-situ measurements in all seasons: Quantifying and reducing errors using dynamic grey-box methods. Energy Build. **167**, 290–300 (2018). https://doi.org/10.1016/j.enbuild.2018.02.048
50. E. Genova, G. Fatta, The thermal performances of historic masonry: In-situ measurements of thermal conductance on calcarenite stone walls in Palermo. Energy Build. **168**, 363–373 (2018). https://doi.org/10.1016/j.enbuild.2018.03.009

51. F. Björk, T. Enochsson, Properties of thermal insulation materials during extreme environment changes. Constr. Build. Mater. **23**, 2189–2195 (2009). https://doi.org/10.1016/j.conbuildmat.2008.12.006
52. I.N. Grubeša, M. Teni, H. Krstić, M. Vračević, Influence of freeze/thaw cycles on mechanical and thermal properties of masonry wall and masonry wall materials. Energies **12**, 1–11 (2019). https://doi.org/10.3390/en12081464
53. L. Evangelisti, C. Guattari, F. Asdrubali, Influence of heating systems on thermal transmittance evaluations: Simulations, experimental measurements and data post-processing. Energy Build. **168**, 180–190 (2018). https://doi.org/10.1016/j.enbuild.2018.03.032
54. C. Guattari, L. Evangelisti, P. Gori, F. Asdrubali, Influence of internal heat sources on thermal resistance evaluation through the heat flow metre method. Energy Build. **135**, 187–200 (2017). https://doi.org/10.1016/j.enbuild.2016.11.045
55. P.G. Cesaratto, M. De Carli, A measuring campaign of thermal conductance in situ and possible impacts on net energy demand in buildings. Energy. Build. **59**, 29–36 (2013). https://doi.org/10.1016/j.enbuild.2012.08.036
56. A.H. Deconinck, S. Roels, Comparison of characterisation methods determining the thermal resistance of building components from onsite measurements. Energy Build. **130**, 309–320 (2016). https://doi.org/10.1016/j.enbuild.2016.08.061
57. S.-H. Kim, J.-H. Kim, H.-G. Jeong, K.-D. Song, Reliability field test of the air-surface temperature ratio method for in situ measurement of U-values. Energies **11**, 1–15 (2018). https://doi.org/10.3390/en11040803
58. E. Lucchi, Applications of the infrared thermography in the energy audit of buildings: A review. Renew Sustain Energy Rev. **82**, 3077–3090 (2018). https://doi.org/10.1016/j.rser.2017.10.031
59. E. Barreira, R.M.S.F. Almeida, J.M.P.Q. Delgado, Infrared thermography for assessing moisture related phenomena in building components. Constr. Build. Mater. **110**, 251–269 (2016). https://doi.org/10.1016/j.conbuildmat.2016.02.026
60. M. O'Grady, A.A. Lechowska, A.M. Harte, Infrared thermography technique as an in-situ method of assessing the heat loss through thermal bridging. Energy Build. **135**, 20–32 (2017). https://doi.org/10.1016/j.enbuild.2016.11.039
61. D. Bienvenido-Huertas, J.A.F. Quiñones, J. Moyano, C.E. Rodríguez-Jiménez, Patents analysis of thermal bridges in slab fronts and their effect on energy demand. Energies **11**, 2222 (2018). https://doi.org/10.3390/en11092222
62. A. Sfakianaki, K. Pavlou, M. Santamouris et al., Air tightness measurements of residential houses in Athens, Greece. Build. Environ. **43**, 398–405 (2008). https://doi.org/10.1016/J.BUILDENV.2007.01.006
63. T. Taylor, J. Counsell, S. Gill, Energy efficiency is more than skin deep: Improving construction quality control in new-build housing using thermography. Energy Build. **66**, 222–231 (2013). https://doi.org/10.1016/J.ENBUILD.2013.07.051
64. R. Albatici, A.M. Tonelli, *On Site Evaluation of U-value of Opaque Building Elements: A New Methodology*. In PLEA 2008—25th Conference on Passive and Low Energy Architecture (2008)
65. G. Dall'O', L. Sarto, A. Panza, Infrared screening of residential buildings for energy audit purposes: Results of a field test. Energies **6**, 3859–3878 (2013). https://doi.org/10.3390/en6083859
66. R. Madding, *Finding R-values of Stud-Frame Constructed Houses with IR Thermography*. Proc InfraMation (2008)
67. P.A. Fokaides, S.A. Kalogirou, Application of infrared thermography for the determination of the overall heat transfer coefficient (U-Value) in building envelopes. Appl. Energy **88**, 4358–4365 (2011). https://doi.org/10.1016/j.apenergy.2011.05.014
68. B. Tejedor, M. Casals, M. Gangolells, Assessing the influence of operating conditions and thermophysical properties on the accuracy of in-situ measured U-values using quantitative internal infrared thermography. Energy Build. **171**, 64–75 (2018). https://doi.org/10.1016/j.enbuild.2018.04.011

69. B. Tejedor, M. Casals, M. Gangolells, X. Roca, Quantitative internal infrared thermography for determining in-situ thermal behaviour of walls. Energy Build. **151**, 187–197 (2017). https://doi.org/10.1016/j.enbuild.2017.06.040
70. D. Bienvenido-Huertas, J. Bermúdez, J. Moyano, D. Marín, Comparison of quantitative IRT to estimate U-value using different approximations of ECHTC in multi-leaf walls. Energy Build. **184**, 99–113 (2019). https://doi.org/10.1016/j.enbuild.2018.11.028
71. D. Bienvenido-Huertas, J. Bermúdez, J.J. Moyano, D. Marín, Influence of ICHTC correlations on the thermal characterization of walls using the quantitative internal infrared thermography method. Build. Environ. **149**, 512–525 (2019). https://doi.org/10.1016/j.buildenv.2018.12.056
72. R.L. Earle, M.D. Earle, *Unit Operations in Food Processing* (1983)
73. J.P. Holman, *Heat Transfer*, 6th edn. (McGraw-Hill Inc., New York, 1986)
74. A. Kylili, P.A. Fokaides, P. Christou, S.A. Kalogirou, Infrared thermography (IRT) applications for building diagnostics: A review. Appl. Energy **134**, 531–549 (2014)
75. B. Lehmann, K. Ghazi Wakili, T. Frank et al., Effects of individual climatic parameters on the infrared thermography of buildings. Appl. Energy **110**, 29–43 (2013). https://doi.org/10.1016/j.apenergy.2013.03.066
76. S. Van De Vijver, M. Steeman, N. Van Den Bossche, et al., The Influence of Environmental Parameters on the Thermographic Analysis of the Building Envelope. In QIRT 2014—12th International Conference on Quantitative InfraRed Thermography (2014)
77. International Organization for Standardization, ISO 13786:2017 Thermal Performance of Building Components—Dynamic Thermal Characteristics—Calculation Methods (2017)
78. H.S. Carslaw, J.C. Jaeger, *Conduction of Heat in Solids*. (Oxford Clarendon Press, 1959)
79. G. Pernigotto, A. Prada, F. Patuzzi et al., Characterization of the dynamic thermal properties of the opaque elements through experimental and numerical tests. Energy Procedia. **78**, 3234–3239 (2015). https://doi.org/10.1016/j.egypro.2015.11.786

Chapter 3
Methodological Framework of Artificial Intelligence Algorithms and Generation of the Dataset

3.1 Introduction

The analysis of the state-of-the-art methods to characterize thermal properties has shown the importance of the theoretical methods (both of stationary and periodic properties) and the difficulty to characterize the existing buildings correctly [1]. The high probability in the deviation of the values obtained is because the constructive composition of walls is unknown or the thermal conductivity values are not representative. However, its wide use to justify the regulations of each country [2] and to validate experimental results [3], as well as its importance to estimate the energy performance [4], requires to characterize accurately these thermal properties in the existing buildings. The analysis of new methodologies to estimate these thermal properties accurately is an aspect to be studied as it would remove one of the major barriers to apply the theoretical methods of ISO 6946 and ISO 13786 in the existing buildings. For this purpose, the bases of experimental methods (e.g. HFM or THM) should be used to join the monitoring of the existing wall with the application of theoretical procedures.

Another aspect found in the review of the state of the art is the deviation presented by THM by using a theoretical value for the total heat transfer coefficient. The deviations presented by the results of THM when the theoretical value is used for the total heat transfer coefficient are not adjusted to the environmental conditions, or the emissivity characteristics could limit its use [5]. In view of this circumstance, having new analysis methodologies to characterize the stationary thermal transmittance of HFM would eliminate the errors related to the use of a theoretical value. Thus, providing THM with a greater possibility of approaches to analyse data by using input variables of THM (i.e. the internal surface temperature and the internal and external air temperatures) would imply a greater use of the method similarly to HFM.

For this purpose, the use of artificial intelligence algorithms is proposed to generate predictor models to address these problems. The automatic learning algorithms chosen were multilayer perceptrons and random forests because of both the existing

D. Bienvenido-Huertas and C. Rubio-Bellido, *Optimization of the Characterization of the Thermal Properties of the Building Envelope*, SpringerBriefs in Applied Sciences and Technology, https://doi.org/10.1007/978-3-030-63629-6_3

differences in the characteristics in each algorithm and the good results obtained in various research studies. The following subsections describe the algorithms and indicate the configurable parameters considered in the research. Then, the dataset and the training and testing of the models are described.

3.2 Artificial Intelligence Algorithms

3.2.1 Artificial Neural Networks

Artificial neural networks (ANNs) are statistical models that simulate the neurological brain structure to solve problems, both linear and nonlinear [6]. A biological neural network is a set of interconnected neurons that defines a circuit. Neural networks, as a kind of bio-inspired computation, are based on the hypothesis that it is possible to imitate certain parts of the neurons, such as axons, the cell body or dendrites (Fig. 3.1a), by using simplified statistical models. They are based on the way the information is transmitted to and processed in a biological neuron: input signals go inside the neuron through dendrites and are sent to the axon after receiving enough signals to provide an output signal that could go to other neurons. Signals are weighted through connections, and various levels of importance are given. Thus, the artificial neural network imitates the whole process by using an activation function and through the sum of input signals (Fig. 3.1b).

Multilayer perceptrons (MLPs) are the ANN architecture providing the best features as they are models supervised with the capability of universal approximation [7–9] to address complex problems, both classification [10] and regression [11]. MLPs are characterized by having an architecture of three or more layers (Fig. 3.2): an input layer, one or several intermediate layers (known as hidden layers), and an output layer. The input layer corresponds to the input of the values of the input variables of the dataset used, and the output layer corresponds to the response given by the model. There are some neurons in each layer of a perceptron. The output value

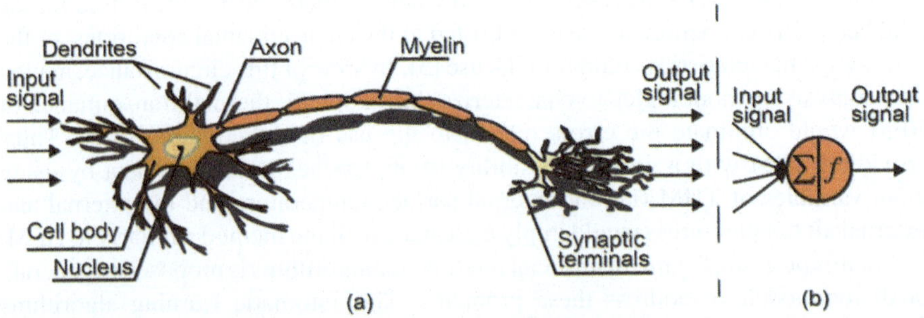

Fig. 3.1 Scheme of neurons: **a** a biological neuron and **b** an artificial neuron

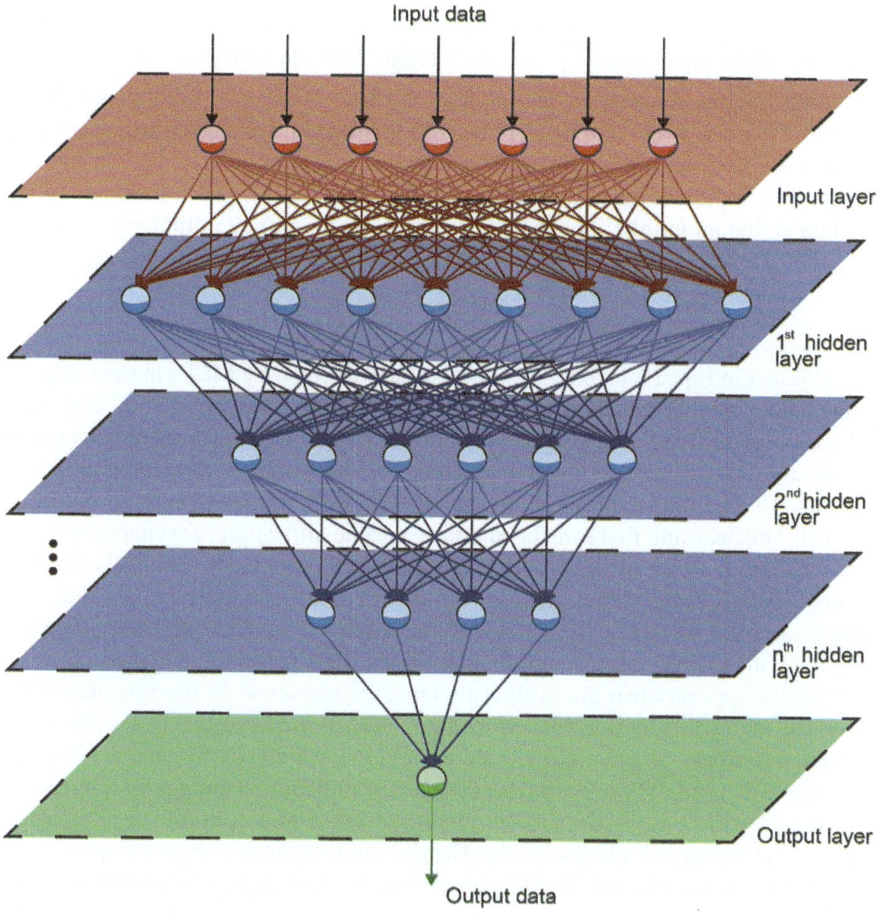

Fig. 3.2 Scheme of an MLP with various hidden layers

of each neuron is obtained by summing the values of the input neurons weighted by synaptic weights and applying an activation function (Eq. (3.1)). These connections spread towards the output layer (Eq. (3.2)), obtaining the response value estimated by the system ($\widehat{Y}_{MLP}$).

$$y_k = \sigma\left(\sum_{j=1}^{d} w_{kj}^{(1)} x_j + w_{k0}^{(1)} x_0\right) \tag{3.1}$$

$$\widehat{Y}_{MLP} = \sigma\left(\sum_{k=1}^{M} w_{lk}^{(2)} \sigma\left(\sum_{j=0}^{d} w_{kj}^{(1)} x_j\right) + w_{l0}^{(2)} y_0\right) \tag{3.2}$$

where x_j are the inputs of the system, $w_{k0}^{(1)}$ and x_0 are the weight and the input of the bias node of the input layer, $w_{kj}^{(1)}$ are the weights of the hidden layer, $w_{l0}^{(2)}$ and y_0 are the weight and the input of the bias node of the hidden layer, $w_{lk}^{(2)}$ are the weights of the output layer,y_k is the output of a node of the hidden layer, and σ is the activation function. In the research conducted by using MLPs, a sigmoidal activation function was used in both the hidden and the output layer (Eq. (3.3)). The advantage of this function is that an infinite input set is compressed into a finite output set.

$$\sigma = \frac{1}{1 + e^{-x}} \tag{3.3}$$

As shown in Eqs. (3.1) and (3.2), the estimation given by MLPs mainly depends on the values assigned to the synaptic weights. The main objective of the algorithm is therefore adjusting these weights to guarantee the greatest adjustment between the estimated output value and the actual value. For this purpose, a learning algorithm is applied to a training dataset. In this case, MLPs were trained by retro-propagation [12–14]. This learning first selects randomly a value allocation for synaptic weights, then includes a series of data randomly introduced, and finally analyses the error obtained between the output value of the model and the actual output value. This analysis is used to adjust the values of the synaptic weights, and the process is repeated with various instances of the dataset until the process is finished.

Thus, the appropriate training algorithm must be chosen in the training of the MLP to reduce the error function related to the model. For this research, the Broyden–Fletcher–Goldfarb–Shanno (BFGS) algorithm [15], which belongs to quasi-Newton methods, was used. These methods present a significant advantage in comparison with the Newton's methods as they are more rapid and consume less resources because the direct calculation of the Hessian and its inverse is not required. For this purpose, approaches of the Hessian inverse are calculated in each iteration.

Likewise, architectures play an important role to determine the MLP with the best estimations, so the number of nodes and hidden layers that best adjust estimated output values should be determined. For this reason, in the several applications of MLPs in this research, the number of nodes varied in the hidden layers.

To conclude, it is worth emphasizing the level of importance recently acquired by ANNs through MLPs in the fields of building, energy efficiency, and thermal characterization. Various studies related to the building energy analysis are as follows:

- Magalhães et al. [16] assessed the use of an ANN to predict the heating energy demand according to the occupants' behaviour.
- Deb et al. [17] used an ANN to predict the energy saving related to the use of HVAC systems in office buildings, and the results with multiple linear regressions were compared.
- Buratti et al. [18] developed an ANN to estimate the thermal transmittance of wooden windows through simple parameters (e.g. the type of window or the thickness of the framework).

- Chudzik [19] assessed the use of an ANN to estimate the heat transfer coefficient in the surface of insulating materials, so that thermal properties could be determined more precisely.
- Mitra et al. [20] assessed an ANN to determine the thermal resistance of cotton materials.
- Aznar et al. [21] designed time-delay neural networks to estimate the temperature value predicted in the next 30 min in each material constituting building walls.

3.2.2 *Random Forest*

Another automatic learning algorithm most applied is tree-type algorithms. The classification and regression tree (CART) algorithm is among the most used. This algorithm was developed by Breiman et al. [22] and is characterized by building models in the form of reverse tree (Fig. 3.3). These trees are made up of internal nodes corresponding to the variables, arches corresponding to the values of the source node, and leaves corresponding to the value of the dependent variable. They are easily understood because their scheme is constituted by nodes and leaves [23]. Thus, the algorithm operates by dividing the input space into subregions, so complex problems are simplified with simple models [24].

In the model development, the optimal structure is obtained by a binary recursive partitioning process of the dataset used by a partitioning rule of each node of the model. This rule is established by reducing the residual sum of squares (Eq. (3.4)). After the induction process, the application of the pruning (i.e. the removal of inefficient nodes) allows the complexity of the model to be generalized and reduced. The depth of the tree and the minimum number of instances per node should be established to configure the algorithm.

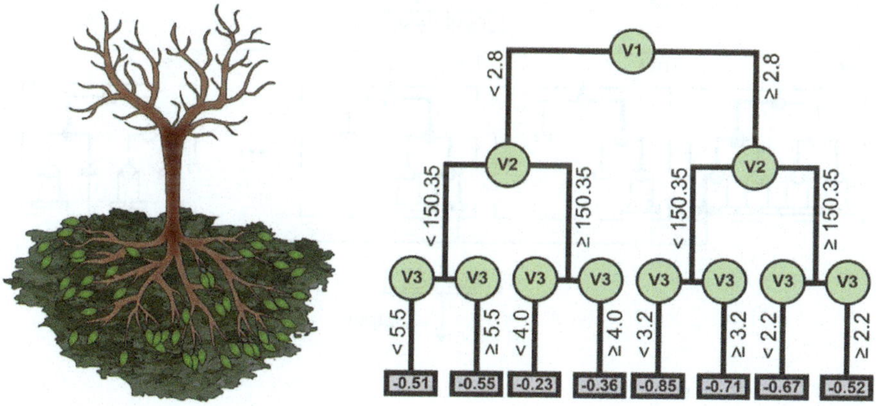

Fig. 3.3 Scheme of a CART model

$$RSS = \sum_{i=1}^{n} (y_i - \hat{y}_i)^2 \tag{3.4}$$

where *RSS* is the residual sum of squares, y_i is the actual output value of each observation *i*, and $\hat{y}_i$ is the value estimated by the model.

Although the CART algorithm has been widely used because its use and understanding are easy, its limitations related to its use have been widely studied [25, 26]. So, the random forest (RF) algorithm is an opportunity to use CART models with a better performance in estimations. RF models are a solution to these problems by generating a set of CART models that are developed in parallel to reduce the variance and the bias of the new model [27, 28]. So, the structure of RF models is a forest of trees (Fig. 3.4). The RF model provides the estimation through the average of the estimations of each CART model:

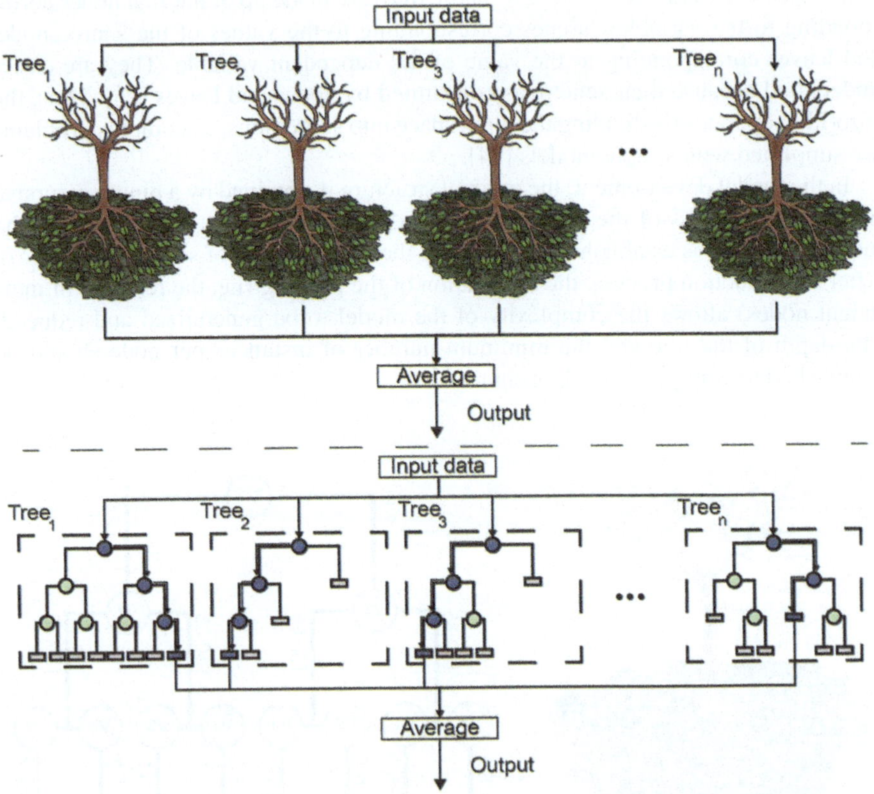

Fig. 3.4 Scheme of an RF model

$$\widehat{Y}_{RF} = \frac{1}{T}\sum_{t=1}^{T}\widehat{Y}_t \quad (3.5)$$

where T is the number of trees, and $\widehat{Y}_t$ is the output value provided by each tree.

Another advantage of RF is the use of large datasets. In addition, they are not affected by atypical values in the instances of the training dataset [29], and their learning is through a learning combined method. RF, as it is a combined learning method (through bootstrap aggregation), obtains a better performance than an individual model [30]. To train RF models, the algorithm first selects N bootstrap samples of the training dataset [28]. Samples are generated by taking random samples of the dataset that guarantee its uniformity (in some cases, some instances could appear repeatedly in various samples). Each bootstrap sample generates a regression tree of the forest. In addition, each node of each tree is divided by using a subset of m predictors randomly selected, so the influence of the strongest predictors is reduced.

Like for MLPs, the architecture or form of the RF model affects its performance. In this regard, the number of trees that constitutes the model significantly influences its performance [31].

Finally, like with ANNs, recent studies on the application of RF in buildings could be emphasized:

- Smarra et al. [32] assessed the use of RF models to use them as predictive control models of buildings to guarantee an energy saving and an appropriate thermal comfort. The application of the methodology developed by these authors with three types of data (historic data, simulated data, and data from a case study) guaranteed the robustness of the approach with RF.
- Wang et al. [33] applied RF to estimate the hourly electricity consumption in educational buildings in Italy. The estimations of the RF models were compared to those of the support vector regression models and CART models. The results showed a better prediction obtained by RF models, with a percentage increase between 5 and 25% in the performance index.
- Lu et al. [34] used RF to assess the hourly performance in geothermal heat pumps, and the estimations were analysed with those obtained by ANN. The results showed a better performance for the RF models not just as regards the degree of adjustment but also as regards the interpret ability of the model and the computational cost for its development.

3.3 Dataset Design

One of the main requirements to apply artificial intelligence algorithms appropriately is having a large dataset to guarantee the correct learning of the models developed. Much time is required to obtain a large test sample, so this could limit the use of

Fig. 3.5 Photograph of some of the buildings analysed

an approach with these characteristics. For this reason, two-dimensional transitory simulations were designed to generate the datasets required.

First, an experimental campaign was conducted with HFM and THM in 30 building walls in Spain (Fig. 3.5). Some walls were used to several monitorings, so the number of tests was greater than 30.

The list of the equipment included in Table 3.1 was used for the tests. The equipment was placed by following the criteria and recommendations indicated in both ISO 9869-1 and the research studies on HFM and THM. The heat flux plate and thermocouples were placed far from singular points or thermal bridges to avoid two-dimensional and three-dimensional effects in the heat flux (the infrared camera eased this aspect in some walls). To place the heat flux plate, a thin silicon grease layer was applied, and the thermocouples were fixed to the walls by mastics with a high thermal conductivity. Mastics were placed 150 cm above the floor, at 10 cm between them and at 2 cm from the mortar joints of the internal layer of the wall (Fig. 3.6a). The air temperature sensors were horizontally aligned and at 30 cm from the wall to avoid convective effects (Fig. 3.6b). Regarding the data logger TESTO 435-2, in the connection of the surface temperature probes with the data logger, there was another temperature probe to measure the internal air temperature (Fig. 3.6c). Thus, the location of this data logger influenced the measurement of the internal temperature, so it was guaranteed that the equipment was placed at a similar height to that of the remaining probes. Tests were performed in various seasons of the year. The duration of each test ranged between 72 and 168 h, according to several factors such as the possibility of having the equipment placed or the possible impact of weather agents (e.g. raining forecast during tests).

Table 3.1 List of the equipment used in the experimental campaigns

Equipment/probe	Variable	Measurement range	Accuracy
ALMEMO 2590-4AS data logger			
T 190-3 probe	Temperature	−10–105 °C	± 0.05 K ± 0.05%
FQA018C probe	Heat flux	± 2000 W/m^2	5%
TESTO 435-2 data logger			
0614 1635 probe	Temperature	−20–70 °C	± 0.1 °C
0632 9735 probe	Temperature	−20–70 °C	± 0.3 °C

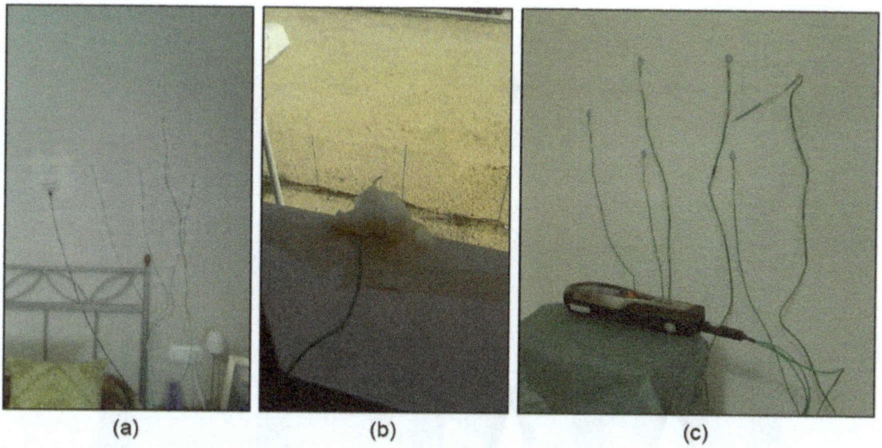

Fig. 3.6 Examples to place equipment and probes: **a** the alignment of the surface temperature probes; **b** the placing of the external temperature probe; and **c** the placing of the TESTO 435-2 data logger

These measurements obtained 163 time series. The simulations were based on a combination of these measurements with simulated wall models (Fig. 3.7). For this purpose, 140 wall typologies were modelled. The design of these typologies was based on both the walls included in the Constructive Elements Catalogue of the Spanish Building Technical Code [35] and those from several cataloguing studies of walls that belong to the Spanish building stock [36, 37]. Therefore, the designed models correspond to wall typologies of the typical building periods of the Spanish building stock. Regarding the walls built in building periods when a regulation on energy efficiency was available, there were several insulating materials, including the extruded polystyrene, the polyurethane, the mineral wool, and the expanded polystyrene. There materials were selected as they are the most usual insulation typologies in building designs [38]. As the walls were designed in the simulation process, their layers were accurately known (i.e. the type of material, the thickness, and the thermal properties of each layer). Thus, there was a dataset whose thermal properties of ISO 6946 and ISO 13786 of each simulated wall were accurately known.

Each wall typology was combined with a set of 163 time series of indoor and outdoor air temperature obtained in in situ measurements. A total of 22,820 combinations between the wall typologies and the time series was obtained. As mentioned above, measurements were conducted under several test conditions. As a result, there were monitorings under favourable and unfavourable conditions, thus encompassing a wide variety test conditions.

Two-dimensional transitory simulations were conducted in each combination to obtain the variables of surface temperature and heat flux (Fig. 3.8). Other aspects related to the simulation process were the time of data acquisition and the location of the probe of the surface temperature and the heat flux. On the one hand, intervals of 15 min were applied, which coincided with those used in the measurement of

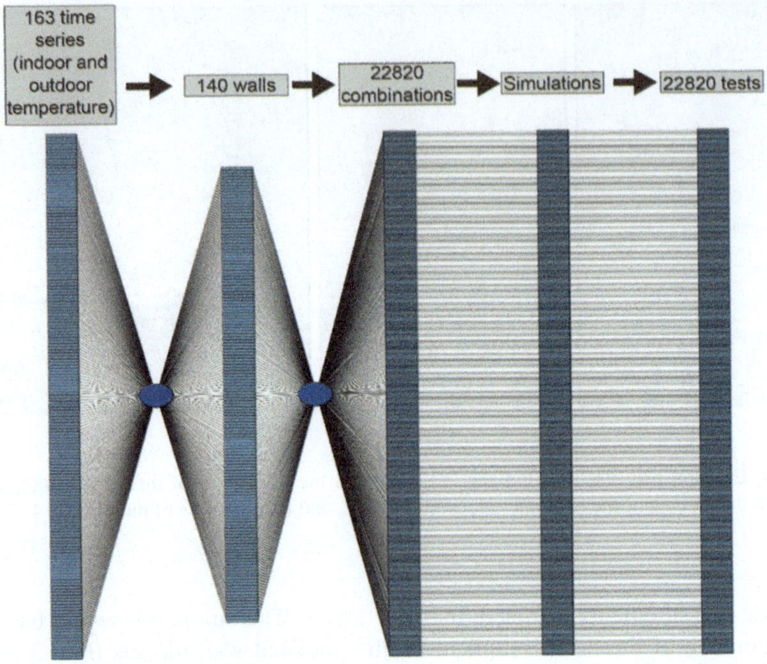

Fig. 3.7 Workflow of the simulation process

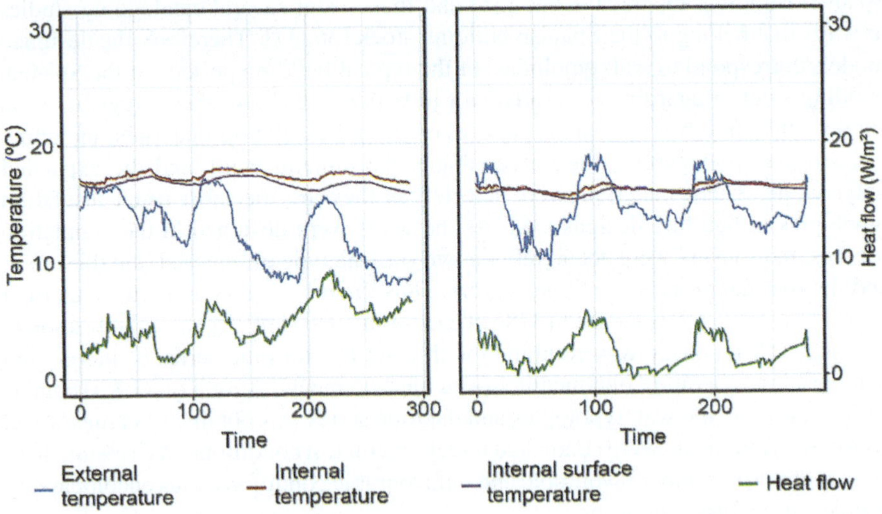

Fig. 3.8 Example of two simulated time series

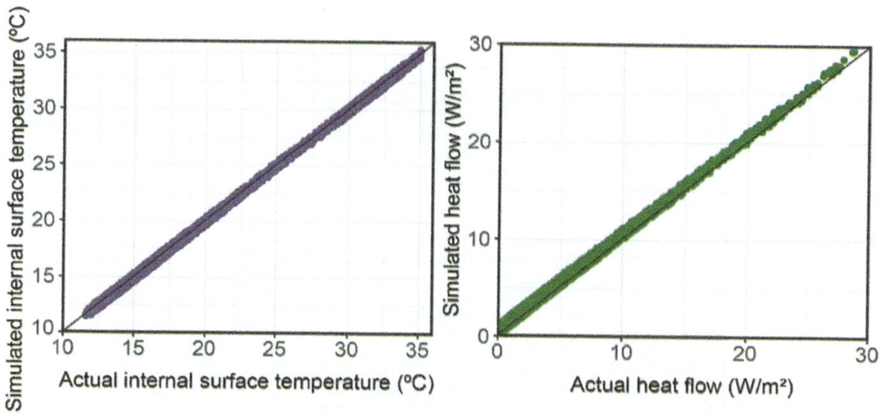

Fig. 3.9 Point clouds between the simulated and actual values of the surface temperature and the heat flux

the indoor and outdoor air temperatures. On the other hand, the measured variables were located 150 cm above the floor to avoid two-dimensional effects in the heat flux because of the junction with the slab. This criterion is in accordance with the recommendations to perform HFM and THM tests.

Actual monitoring data were used, so the simulated data were validated, thus guaranteeing the representation of the data used. Figure 3.9 represents the point clouds between the simulated and actual values. In both cases, the simulated and actual values were appropriately approached. In line with this aspect, the absolute mean error was 0.14 °C for the internal surface temperature and 0.23 W/m^2 for the heat flux, thus guaranteeing that the simulated values were close to those obtained under actual conditions and that the results of the simulated tests were assimilated to actual tests performed under the same conditions.

3.4 Training and Testing Processes

The dataset developed was used to train and test models. However, the number of instances for training and testing was different. Thus, the dataset was randomly divided as follows: 75% for the training and 25% for the testing.

To train the models, a tenfold cross-validation was used (Fig. 3.10). The advantages of using the cross-validation to train the models were significant as the bias and the variance of the models could be reduced [39]. All training sets were randomly divided into 10 subsets. For each iteration, nine subsets were used for the training,

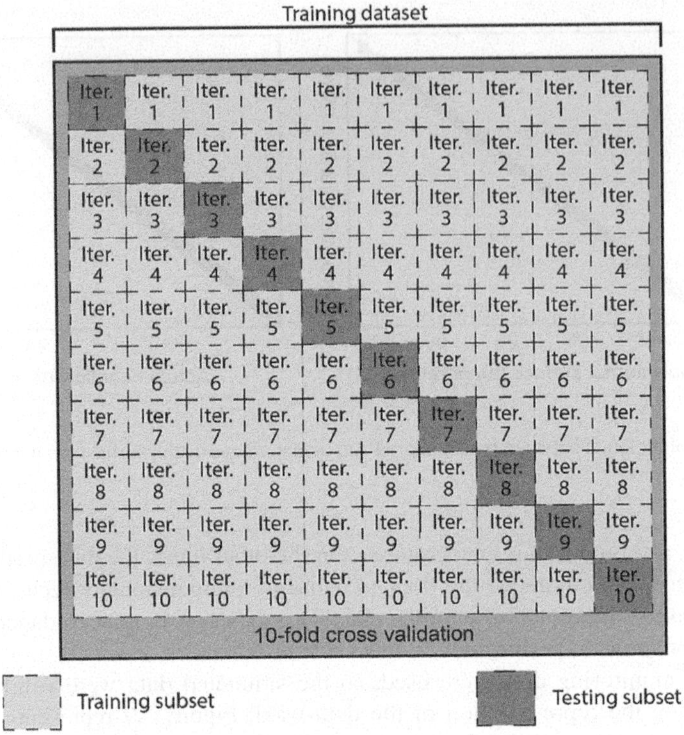

Fig. 3.10 Scheme of the tenfold cross-validation in a training dataset

and the remaining subset for the testing. The performance of the models was obtained from the average value of 10 iterations.

As mentioned above, the appropriate configuration of the MLP and RF models influences their performance. For this reason, various architectures were assessed for the two algorithms:

- Regarding the MLP models, the number of nodes varied between 2 and 15 in the hidden layer until determining the optimal configuration.
- Regarding the RF models, the number of nodes varied between 2 and 50 until determining the optimal configuration.

To assess the performance of the models in the training and testing phases, various statistical parameters were used. As the approaches varied throughout the research (regression and classification approaches were distinguished), the statistical parameters changed according to the type of the output variable. Regarding the regression approaches, the quality indicators were as follows: the determination coefficient (R^2) (Eq. (3.6)), the mean absolute error (MAE) (Eq. (3.7)), and the root mean square error (*RMSE*) (Eq. (3.8)). These parameters have been widely used to assess the performance of regression models.

$$R^2 = 100\left(1 - \frac{\sum_{i=1}^{n}(r_i - e_i)^2}{\sum_{i=1}^{n}(r_i - \bar{e}_i)^2}\right) \tag{3.6}$$

$$MAE = \frac{\sum_{i=1}^{n}|r_i - e_i|}{n} \tag{3.7}$$

$$RMSE = \left(\frac{\sum_{i=1}^{n}(r_i - e_i)^2}{n}\right)^{1/2} \tag{3.8}$$

where r_i is the actual value of the output variable, e_i is the estimated value of the output variable, and n is the number of observations of the dataset used.

Regarding the classification approach, the statistical parameters analysed were as follows: the true positive ratio (TP) (Eq. (3.9)), the false positive ratio (FP) (Eq. (3.10), the kappa statistic of Cohen (Eq. (3.11)), and the area under the receiver operating characteristic (ROC) curve (Eq. (3.12)). TP and FP ratios indicate the success percentage in the estimations performed by the model, the kappa statistic determines the coincidence of the estimation with the actual class, and the area under the ROC curve determines the probability that the model correctly classifies the class analysed, existing a different value for each label or output.

$$\text{TP} = \frac{\text{Number of correctly classified observations}}{\text{Total number of observations}} \tag{3.9}$$

$$\text{FP} = \frac{\text{Number of incorrectly classified observations}}{\text{Total number of observations}} \tag{3.10}$$

$$K = \frac{p_o - p_e}{1 - p_e} \tag{3.11}$$

$$\text{AreaROC} = \int_0^1 \left(1 - G\left(F^{-1}(1 - t)\right)\right)dt \tag{3.12}$$

where p_o is the relative observed agreement between observers, p_e is the hypothetical probability of agreement by chance, and G and F are the distribution of positive and negative samples, respectively.

References

1. G. Ficco, F. Iannetta, E. Ianniello et al., U-value in situ measurement for energy diagnosis of existing buildings. Energ. Build. **104**, 108–121 (2015). https://doi.org/10.1016/j.enbuild.2015.06.071
2. B. Rodríguez-Soria, J. Domínguez-Hernández, J.M. Pérez-Bella, J.J. Coz-Díaz, Review of international regulations governing the thermal insulation requirements of residential buildings and the harmonization of envelope energy loss. Renew Sustain Energ. Rev. **34**, 78–90 (2014). https://doi.org/10.1016/j.rser.2014.03.009
3. International Organization for Standardization, ISO 9869-1:2014—Thermal insulation—Building elements—In situ measurement of thermal resistance and thermal transmittance. Part 1: Heat flow meter method. (Geneva, Switzerland, 2014)
4. D. Bienvenido-Huertas, C. Rubio-Bellido, JA Pulido-Arcas, A Pérez-Fargallo (2020) Towards the implementation of periodic thermal transmittance in Spanish building energy regulation. J. Build. Eng. **31** https://doi.org/10.1016/j.jobe.2020.101402
5. D. Bienvenido-Huertas, J. Moyano, D. Marín, R. Fresco-Contreras, Review of in situ methods for assessing the thermal transmittance of walls. Renew. Sustain. Energ. Rev. **102**, 356–371 (2019). https://doi.org/10.1016/j.rser.2018.12.016
6. S.S. Haykin, *Neural Networks and Learning Machines*. Pearson Upper Saddle River (2009)
7. A.R. Barron, Universal approximation bounds for superpositions of a sigmoidal function. IEEE Trans. Inf. Theor. **39**, 930–945 (1993)
8. G. Cybenko, Approximation by superpositions of a sigmoidal function. Math. Control Signals Syst. **2**, 303–314 (1989)
9. K. Hornik, M. Stinchcombe, H. White, Multilayer feedforward networks are universal approximators. Neural Netw. **2**, 359–366 (1989). https://doi.org/10.1016/0893-6080(89)90020-8
10. S. Raghu, N. Sriraam, Optimal configuration of multilayer perceptron neural network classifier for recognition of intracranial epileptic seizures. Expert Syst. Appl. **89**, 205–221 (2017). https://doi.org/10.1016/j.eswa.2017.07.029
11. L. Mba, P. Meukam, A. Kemajou, Application of artificial neural network for predicting hourly indoor air temperature and relative humidity in modern building in humid region. Energ. Build. **121**, 32–42 (2016). https://doi.org/10.1016/j.enbuild.2016.03.046
12. D.E. Rumelhart, G.E. Hinton, R.J. Williams, Learning representations by back-propagating errors. Nature **323**, 533–536 (1986). https://doi.org/10.1038/323533a0
13. Y.N. Wang, A neural network adaptive control based on rapid learning method and application. Adv. Molding Anal. **46**, 27–34 (1994)
14. P. Werbos, *Beyond Regression: New Tools for Prediction and Analysis in the Behavior Science*. (Harvard University, 1974)
15. R. Fletcher, *Practical Methods of Optimization* (Wiley, Chichester, New York, Brisbane, Toronto, United States, 1980)
16. S.M.C. Magalhães, V.M.S. Leal, I.M. Horta, Modelling the relationship between heating energy use and indoor temperatures in residential buildings through artificial neural networks considering occupant behavior. Energ. Build. **151**, 332–343 (2017). https://doi.org/10.1016/j.enbuild.2017.06.076
17. C. Deb, S. Eang, M. Santamouris, Using artificial neural networks to assess HVAC related energy saving in retrofitted office buildings. **163**, 32–44 (2018). https://doi.org/10.1016/j.solener.2018.01.075
18. C. Buratti, L. Barelli, E. Moretti, Application of artificial neural network to predict thermal transmittance of wooden windows. Appl. Energ. **98**, 425–432 (2012). https://doi.org/10.1016/j.apenergy.2012.04.004
19. S. Chudzik, Applying infrared measurements in a measuring system for determining thermal parameters of thermal insulation materials. Infrared Phys. Technol. **81**, 296–304 (2017). https://doi.org/10.1016/j.infrared.2016.12.025

20. A. Mitra, A. Majumdar, P.K. Majumdar, D. Bannerjee, Predicting thermal resistance of cotton fabrics by artificial neural network model. Exp. Therm. Fluid. Sci. **50**, 172–177 (2013). https://doi.org/10.1016/j.expthermflusci.2013.06.006
21. F. Aznar, V. Echarri, C. Rizo, R. Rizo, Modelling the thermal behaviour of a building facade using deep learning. PLoS ONE **13**, 1–20 (2018). https://doi.org/10.1371/journal.pone.0207616
22. L. Breiman, J. Friedman, C.J. Stone, R.A. Olshen, Classification and regression trees. Routledge (2017)
23. M. Xu, P. Watanachaturaporn, P.K. Varshney, M.K. Arora, Decision tree regression for soft classification of remote sensing data. Remote Sens. Environ. **97**, 322–336 (2005). https://doi.org/10.1016/j.rse.2005.05.008
24. W. Sun, River ice breakup timing prediction through stacking multi-type model trees. Sci. Total Environ. **644**, 1190–1200 (2018). https://doi.org/10.1016/j.scitotenv.2018.07.001
25. S. Dudoit, J. Fridlyand, T.P. Speed, Comparison of discrimination methods for the classification of tumors using gene expression data. J. Am. Stat. Assoc. **97**, 77–87 (2002)
26. B. Larivière, D. Van Den Poel, Predicting customer retention and profitability by using Random forests and regression forests techniques. Expert Syst. Appl. **29**, 472–484 (2005). https://doi.org/10.1016/j.eswa.2005.04.043
27. L. Breiman, Bagging predictors. Mach. Learn. **24**, 123–140 (1996)
28. L. Breiman, Random forests. Mach. Learn. **45**, 5–32 (2001). https://doi.org/10.1023/A:1010933404324
29. D. Assouline, N. Mohajeri, J.L. Scartezzini, Large-scale rooftop solar photovoltaic technical potential estimation using Random forests. Appl. Energ. **217**, 189–211 (2018). https://doi.org/10.1016/j.apenergy.2018.02.118
30. T.G. Dietterich, Experimental comparison of three methods for constructing ensembles of decision trees: Bagging, boosting, and randomization. Mach. Learn. **40**, 139–157 (2000). https://doi.org/10.1023/A:1007607513941
31. Y. Zhou, G. Qiu, Random forest for label ranking. Expert Syst. Appl. **112**, 99–109 (2018). https://doi.org/10.1016/j.eswa.2018.06.036
32. F. Smarra, A. Jain, T. de Rubeis et al., Data-driven model predictive control using Random forests for building energy optimization and climate control. Appl. Energ. **226**, 1252–1272 (2018). https://doi.org/10.1016/j.apenergy.2018.02.126
33. Z. Wang, Y. Wang, R. Zeng et al., Random forest based hourly building energy prediction. Energ. Build. **171**, 11–25 (2018). https://doi.org/10.1016/j.enbuild.2018.04.008
34. S. Lu, Q. Li, L. Bai, R. Wang, Performance predictions of ground source heat pump system based on Random forest and back propagation neural network models. Energ. Convers Manag. **197**, 111864 (2019). https://doi.org/10.1016/j.enconman.2019.111864
35. Eduardo Torroja Institute for Construction Science, *Constructive Elements Catalogue of the CTE* (2010)
36. S. Domínguez-Amarillo, J.J. Sendra, I. Oteiza, *La envolvente térmica de la vivienda social*. El caso de Sevilla, 1939 a 1979. (Editorial CSIC, Madrid, 2016)
37. F. Kurtz, M. Monzón, B. López-Mesa, Energy and acoustics related obsolescence of social housing of Spain's post-war in less favoured urban areas. The case of Zaragoza. Inf la Construcción 67:m021 (2015). https://doi.org/10.3989/ic.14.062
38. S. Schiavoni, F. D'Alessandro, F. Bianchi, F. Asdrubali, Insulation materials for the building sector: A review and comparative analysis. Renew Sustain Energ. Rev. **62**, 988–1011 (2016). https://doi.org/10.1016/j.rser.2016.05.045
39. R. Kohavi, A Study of Cross-Validation and Bootstrap for Accuracy Estimation and Model Selection. In *International Joint Conference on Artificial Intelligence* (1995)

Chapter 4
Estimation of Stationary Thermal Properties with Artificial Intelligence

4.1 Introduction

This chapter aims to develop a mathematical model by using MLP and RF to estimate the thermal transmittance of ISO 6946 (U_{6946}) through in situ measured variables. For this purpose, the scheme of variables of HFM and THM is used. The results of this research aim to determine the thermal transmittance of ISO 6946 through the variables which would be measured with both experimental methods. After obtaining this value, the experimental results obtained with HFM or THM could be validated according to the criterion included in ISO 9869-1 to validate experimental results.

To achieve this goal, the dataset described in Chap. 3 was used. This dataset was made up of 22,820 simulated tests obtained by combining 163 actual monitorings with 140 wall designs. Each test contributed with an observation in the dataset. Regarding the design of the input variables, two approaches were designed according to the type of in situ method used. Regarding HFM approach, the monitored variables (the internal temperature, the external temperature, and the heat flux), the thickness, and the test duration were considered. Regarding THM approach, the same variables were considered but replacing the heat flux by the internal surface temperature of the wall. Table 4.1 summarizes the input and output variables considered in this research.

This dataset was randomly divided as follows: 75% for the training and 25% for the testing of the models. The statistical parameters were R^2, *MAE*, and *RMSE*. With these parameters, the most appropriate configuration of the MLP and RF models was determined in the training, and then, they were used to assess the performance of the estimations conducted with the testing dataset. Likewise, three monitored walls were individually analysed. These walls were useful to assess in detail the accuracy of the estimations in actual cases. Table 4.2 indicates the thermal properties of these walls and their stationary thermal transmittance value.

D. Bienvenido-Huertas and C. Rubio-Bellido, *Optimization of the Characterization of the Thermal Properties of the Building Envelope*, SpringerBriefs in Applied Sciences and Technology, https://doi.org/10.1007/978-3-030-63629-6_4

Table 4.1 Input and output variables used in each approach

Approach	Input variables	Output variables
HFM	$\overline{T}_{int}$, max(T_{int}), min(T_{int}), $\overline{T}_{ext}$, max(T_{ext}), min(T_{ext}), $\overline{q}$, max(q), min(q), thickness, time	U_{6946}
THM	$\overline{T}_{int}$, max(T_{int}), min(T_{int}), $\overline{T}_{ext}$, max(T_{ext}), min(T_{ext}), $\overline{T}_{s,int}$, max($T_{s,int}$), min($T_{s,int}$), thickness, time	U_{6946}

Table 4.2 Thermophysical properties of the walls selected to be individually analysed

Wall	Layer	Thickness [m]	Thermal conductivity (W/(mK))	Thermal transmittance (W/(m^{2}K))
A	Cement mortar	0.015	1.000	1.31
	Solid brick	0.115	0.850	
	Cement mortar	0.015	1.000	
	Air gap	0.100	–	
	Hollow brick	0.070	0.320	
	Gypsum plaster	0.015	0.570	
B	Cement mortar	0.015	1.000	0.69
	Perforated brick	0.115	0.350	
	Cement mortar	0.015	1.000	
	Air gap	0.010	–	
	Insulation	0.020	0.038	
	Hollow brick	0.070	0.320	
	Gypsum plaster	0.015	0.570	
C	Cement mortar	0.015	1.000	0.57
	Perforated brick	0.115	0.350	
	Cement mortar	0.015	1.000	
	Air gap	0.005	–	
	Insulation	0.040	0.038	
	Laminated plasterboard	0.015	0.250	

4.2 Estimation Performance of the Stationary Thermal Properties Obtained with Artificial Intelligence

The following subsections present the results obtained with the two algorithms (MLP and RF) in the training and testing to estimate the stationary thermal transmittance.

4.2.1 Artificial Neural Network

Regarding the ANNs, the most appropriate architecture of the MLPs was first determined. For this purpose, the architecture of the MLPs was assessed by varying the number of nodes between 2 and 15 until determining the most appropriate configuration. This analysis was performed by assessing the statistical parameters indicated above and distinguishing the most appropriate architecture in each approach. Figure 4.1 shows the values obtained by each architecture in the training. The most appropriate architectures of the MLPs were those of 14 nodes with HFM and 12 nodes with THM. In these configurations, R^2 obtained the highest value, and *MAE* and *RMSE* obtained the lowest values. Any number of neurons greater than five

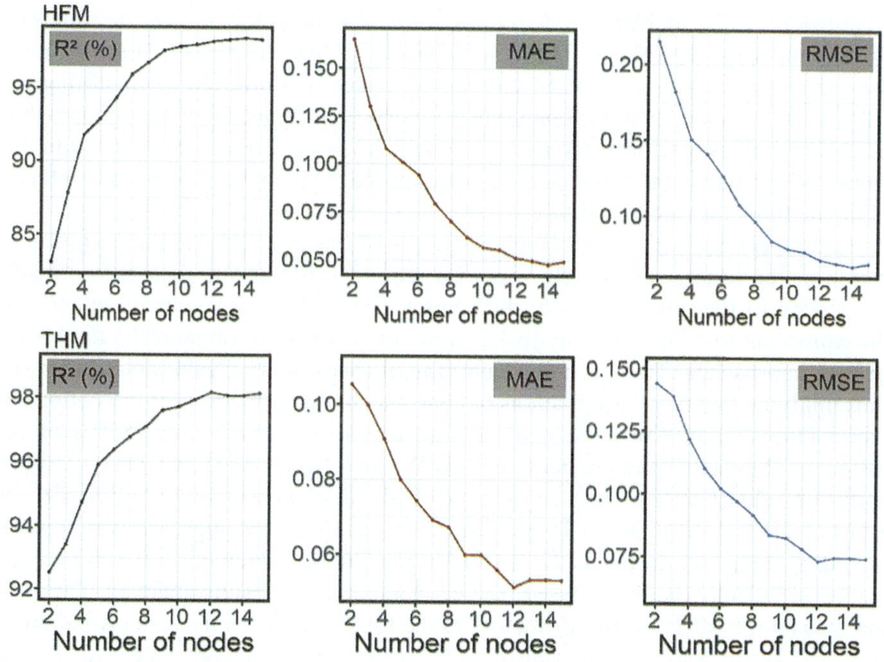

Fig. 4.1 Evaluation of R^2, *MAE*, and *RMSE* in the training phase of the MLP models for the two approaches of in situ variables

Table 4.3 Values obtained in the statistical parameters in the testing with the optimal configuration of the MLP models

Approach	R^2 (%)	*MAE*	*RMSE*
HFM	98.86	0.0407	0.0604
THM	98.52	0.0497	0.0699

nodes in the hidden layer obtained valid results. Regarding HFM, with five or more neurons, values for R^2 greater than 92.89% were obtained, as well as values for *MAE* and *RMSE* lower than 0.1008 and 0.1408, respectively. Regarding THM, these results in the performance in the training were similar, with values for the determination coefficient greater than 95.88%, and with values lower than 0.0799 for *MAE* and lower than 0.1102 for *RMSE*. Thus, the results obtained in the training phase showed that the models had an appropriate degree of adjustment for the input data, so the design of the input variables was appropriate to estimate the output variable.

However, it is crucial to know the performance of the models when estimations are carried out with observations not belonging to the input dataset (i.e. their performance should be known when the results are analysed in new tests). In these cases, the performance of MLPs was appropriate. Table 4.3 includes the results obtained for the statistical parameters. In the testing, the values of the statistical parameters improved in comparison with those obtained in the training. In this regard, the values of R^2 went from 98.17% and 98.42% to values of 98.52% and 98.86%, and the values of *MAE* went from 0.0479 and 0.0512 to values of 0.0407 and 0.0497. Furthermore, the order of the approaches with better performance was inverted in the testing, so HFM went from the approach obtaining the worst performance in the training to that obtaining the best performance in the testing. The small differences detected in the values of the statistical parameters of both approaches emphasized the advantages of the methodology designed.

This aspect could be seen in the histograms of the error obtained in each observation of the testing dataset. By way of a reminder, the testing dataset was made up of 25% of the observations of the dataset designed in the research, which was the equivalent to 5705 tests. Figure 4.2 shows the histograms obtained by the MLPs of both approaches. There is a concentration of cases in error values close to zero, thus implying that most observations obtained accurate estimations of the stationary thermal transmittance with the MLPs. In this regard, only 26.82% and 35.71% of the instances obtained errors greater than 0.05 W/(m^2K) with HFM and THM, respectively. Less than 0.01% of the data analysed obtained errors of up to 0.324 W/(m^2K) in HFM and 0.397 W/(m^2K) in THM. The effectiveness of this approach was corroborated by analysing in detail the thermal transmittance obtained in the three case studies to be individually analysed. Table 4.4 includes the results obtained. The estimations by the MLPs were satisfactory in the three walls, mainly with errors lower than 0.05 W/(m^2K), except in walls A and C with HFM, which obtained error values of up to 0.09 W/(m^2K). Given the low error values obtained with the two models,

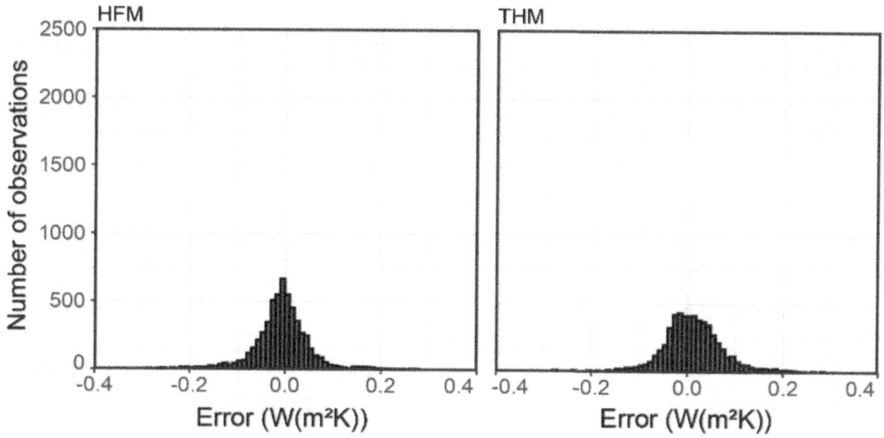

Fig. 4.2 Histograms with the errors obtained by the MLP models in the estimations of the stationary thermal transmittance of the instances of the testing dataset. The histogram is represented with intervals of 0.01 $W/(m^2K)$

Table 4.4 Stationary thermal transmittance results obtained by the MLP models in the three walls to be individually analysed

Wall	U_{6946} $(W/(m^2K))$	HFM		THM	
		Predicted U-value $(W/(m^2K))$	Error $(W/(m^2K))$	Predicted U-value $(W/(m^2K))$	Error $(W/(m^2K))$
A	1.31	1.38	0.07	1.35	0.04
B	0.69	0.66	0.03	0.67	0.02
C	0.57	0.66	0.09	0.61	0.04

the models designed with MLP could be an opportunity to estimate the stationary thermal transmittance appropriately.

4.2.2 Random Forest

Regarding RF, the process was like that of MLPs. First, the most appropriate configuration was determined for the models of both approaches. In this case, the parameter that influences the model performance is the number of trees. The analysis varied the number of trees between 2 and 50. Figure 4.3 shows the performance results obtained in the training with the variations in the number of trees. The optimal number of trees was 46 for HFM and 42 for THM. With these configurations, the RF model of the HFM approach obtained a R^2 of 99.84%, a *MAE* of 0.010, and a *RMSE* of 0.0217, and the THM approach obtained a R^2 of 99.56%, a *MAE* of 0.0182, and a *RMSE*

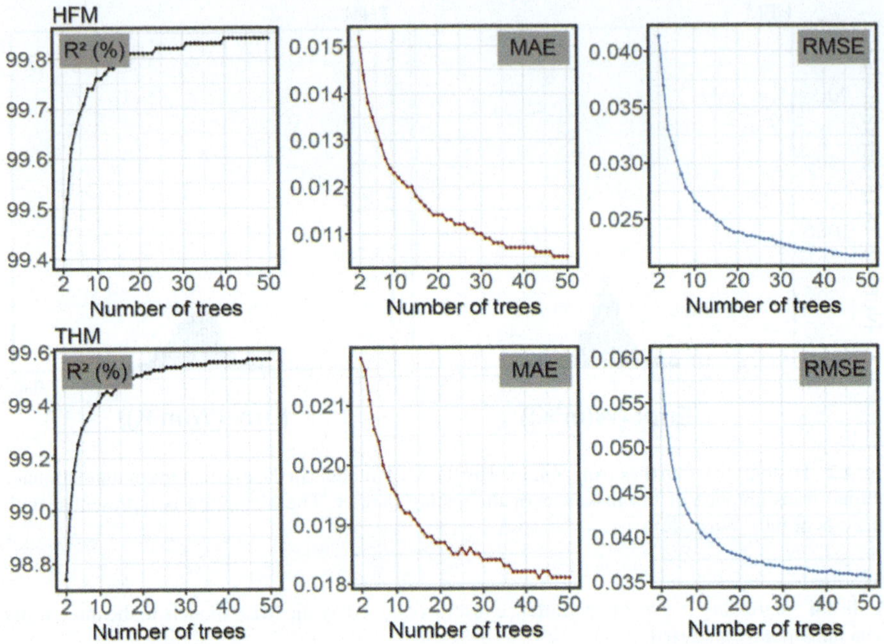

Fig. 4.3 Evaluation of R^2, *MAE*, and *RMSE* in the training of the RF models for the two approaches of in situ variables

of 0.036. However, the performance obtained by the RF models was appropriate in all tree configurations. The reason was that tree configurations greater than 30 trees obtained performances very similar to the optimal configurations for HFM and THM.

Regarding the testing, Table 4.5 shows the values obtained in the statistical parameters. The performances obtained by both approaches in the testing were satisfactory. The determination coefficient was greater than 99.50%, and average error values lower than 0.02 W/(m^2K) were obtained. In addition, the HFM approach obtained better estimations. This could be corroborated by seeing the histograms of the errors obtained in the estimation of each instance (Fig. 4.4). In these histograms, there was a high concentration of instances with errors close to zero: 75.14–57.18% of the instances that made up the testing dataset obtained errors lower than 0.01 W/(m^2K) with the HFM and THM approaches, respectively. Some cases obtained high error values, thus showing the limitations of the model, although the low number of

Table 4.5 Values obtained in the statistical parameters in the testing phase with the optimal configuration of the RF models

Approach	R^2 (%)	*MAE*	*RMSE*
HFM	99.86%	0.0098	0.0210
THM	99.55%	0.0179	0.0375

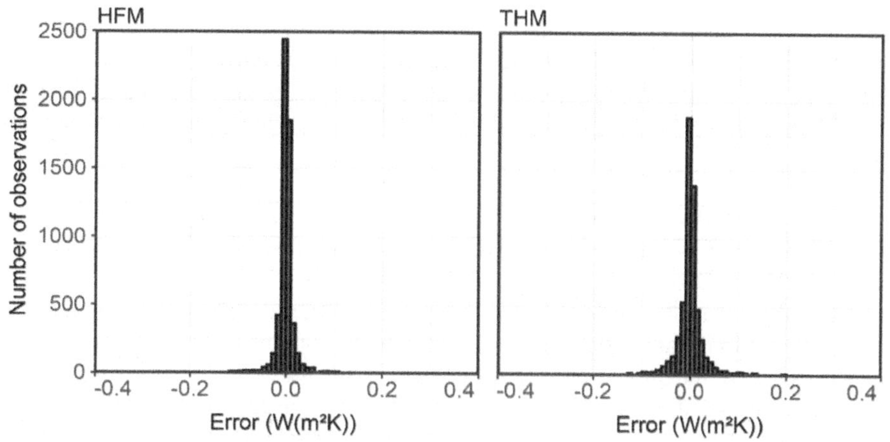

Fig. 4.4 Histograms with the errors obtained with the RF models in the estimations of the stationary thermal transmittance of the instances of the testing dataset. The histogram is represented with intervals of 0.01 W/(m^{2}K)

Table 4.6 Stationary thermal transmittance results obtained with the RF models in the three walls to be individually analysed

Wall	U_{6946} (W/(m^{2}K))	HFM		THM	
		Predicted U-value (W/(m^{2}K))	Error (W/(m^{2}K))	Predicted U-value (W/(m^{2}K))	Error (W/(m^{2}K))
A	1.31	1.30	0.01	1.31	0.00
B	0.69	0.69	0.00	0.69	0.00
C	0.57	0.56	0.01	0.57	0.00

cases (0.01% of the testing dataset) emphasized the reliability of the RF models. Finally, Table 4.6 includes the results obtained in the individual analysis of the three case studies. The estimations conducted by the RF models were satisfactory, with errors lower than 0.01 W/(m^{2}K). Thus, the RF models could be a valid alternative to characterize the stationary thermal transmittance of the existing walls.

4.3 Comparative Analysis

The results of the study showed that the use of MLP and RF models estimated the thermal transmittance value of ISO 6946 accurately. However, what is the most appropriate algorithm according to the results obtained in the previous subsections?

There are differences between the two algorithms that show the greatest potential of RF to estimate the stationary thermal transmittance accurately. In this regard, regarding HFM approach, 75.14% of estimations had an error between 0 W/(m^2K) and 0.01 W/(m^2K) in the RF models, whereas the percentage was lower in MLP (21.27%). This same tendency was shown with the THM approach, as 57.18% of the instances with errors lower than 0.01 W/(m^2K) with the RF model decreased to 12.44% with the MLP model.

In addition, the limit values in which the error of the testing dataset oscillated were lower in the RF models: in HFM, the error was between −0.181 W/(m^2K) and 0.167 W/(m^2K) with RF, and with MLP, the error was between −0.318 W/(m^2K) and 0.324 W/(m^2K)).

Thus, these results show that RF is an appropriate algorithm to estimate the stationary thermal transmittance from ISO 6946. Although MLP had a good performance, RF was the most appropriate algorithm to assess the stationary thermal transmittance because the determination coefficient was greater than 99.55% and more than 5000 tests analysed in the testing phase obtained errors lower than 0.05 W/(m^2K).

Therefore, the results show the possibilities to estimate this variable to automate and avoid errors in the thermal characterization. The use of both MLP and RF would have valid results, although the latter is related to a greater accuracy in estimations. Moreover, the saving of the computing time required to develop the RF models in comparison with the MLP models is emphasized.

Thus, the RF models could be used as an auxiliary evaluation methodology of the thermal transmittance before characterizing the thermal transmittance of the experimental methods. With the reference value obtained through the RF models, it could be determined whether the experimental results are coincident. If there is a deviation greater than 20% between both values, then the test is performed under inappropriate conditions or there is an error in the measurement conducted by the probes.

Chapter 5
Estimating Periodic Thermal Properties with Artificial Intelligence

5.1 Introduction

This chapter aims to develop a mathematical model through MLP and RF to estimate the periodic thermal properties of ISO 13786 through in situ measured variables. In particular, the periodic thermal variables to be estimated are as follows: the periodic thermal transmittance (Y_{12}), the time shift periodic thermal transmittance (φ), the decrement factor (f), the internal thermal admittance (Y_{11}), the time shift internal side (φ_{11}), the external thermal admittance (Y_{22}), the time shift external side (φ_{22}), the internal areal heat capacity (k_1), and the external areal heat capacity (k_2). For this purpose, the scheme of the variables of HFM and THM followed in the research on the stationary thermal transmittance in the previous chapter is used. The results aim to determine the periodic thermal properties of the existing walls through the variables which would be measured with both experimental methods. Thus, engineers and architects would be provided with more accurate information of the performance of the existing buildings to establish more appropriate ECMs.

To achieve this goal, the dataset described in Chap. 3 was used. As the design of the wall was known in the 22,820 tests, the periodic thermal variables could be correctly determined through ISO 13786. Each periodic thermal variable constitutes a different output variable. To address this aspect, independent models were designed for each output variable. Regarding the design of the input variables, two approaches were designed according to the type of monitoring method: HFM or THM. The input and output variables of each approach are included in Table 5.1.

The dataset was randomly divided for the training and testing as follows: 75% of the instances were used for the training, and the remaining 25% for the testing. As they were numeric output variables, the model performance was analysed according to R^2, MAE, and $RMSE$. Likewise, three monitored walls were individually analysed. These walls were useful to assess in detail the accuracy of the estimations in actual cases. Table 5.2 indicates the thermal properties of these walls and their periodic

D. Bienvenido-Huertas and C. Rubio-Bellido, *Optimization of the Characterization of the Thermal Properties of the Building Envelope*, SpringerBriefs in Applied Sciences and Technology, https://doi.org/10.1007/978-3-030-63629-6_5

Table 5.1 Input and output variables used in each approach

Approach	Input variables	Output variables
HFM	$\overline{T}_{int}$, $\max(T_{int})$, $\min(T_{int})$, $\overline{T}_{ext}$, $\max(T_{ext})$, $\min(T_{ext})$, $\overline{q}$, $\max(q)$, $\min(q)$, thickness, time, period	Y_{12}, f, φ, Y_{11}, φ_{11}, Y_{22}, φ_{22}, k_1, k_2
THM	$\overline{T}_{int}$, $\max(T_{int})$, $\min(T_{int})$, $\overline{T}_{ext}$, $\max(T_{ext})$, $\min(T_{ext})$, $\overline{T}_{s,int}$, $\max(T_{s,int})$, $\min(T_{s,int})$, thickness, time, period	

Table 5.2 Thermophysical properties of the walls selected to be individually analysed

Variable	Wall A	Wall B	Wall C
Y_{12}	0.370 W/(m2K)	0.277 W/(m2K)	0.244 W/(m2K)
f	0.283	0.402	0.428
φ	8.506 h	8.364 h	6.552 h
Y_{11}	5.192 W/(m2K)	3.817 W/(m2K)	3.846 W/(m2K)
φ_{11}	1.171 h	2.146 h	2.137
Y_{22}	3.826 W/(m2K)	4.031 W/(m2K)	1.184 W/(m2K)
φ_{22}	2.940 h	3.673 h	3.268 h
k_1	57.657 kJ/(m2K)	59.238 kJ/(m2K)	19.191 kJ/(m2K)
k_2	75.638 kJ/(m2K)	56.035 kJ/(m2K)	55.096 kJ/(m2K)

thermal properties obtained with ISO 13786. These walls were the same used in the research on the stationary thermal transmittance in Chap. 4.

5.2 Estimation Performance of the Periodic Thermal Properties Obtained with Artificial Intelligence

The following subsections show the results obtained with MLP and RF in the training and testing phases to estimate the periodic thermal properties.

5.2.1 Artificial Neural Network

First, the most optimal architecture of the MLP models was determined to estimate each periodic thermal variable. Unlike in Chap. 4, various thermal variables were estimated, so independent models were designed. This meant that the model of each variable was individually analysed to determine the most appropriate architecture. Figures 5.1 and 5.2 show the tendencies obtained in the statistical parameters by

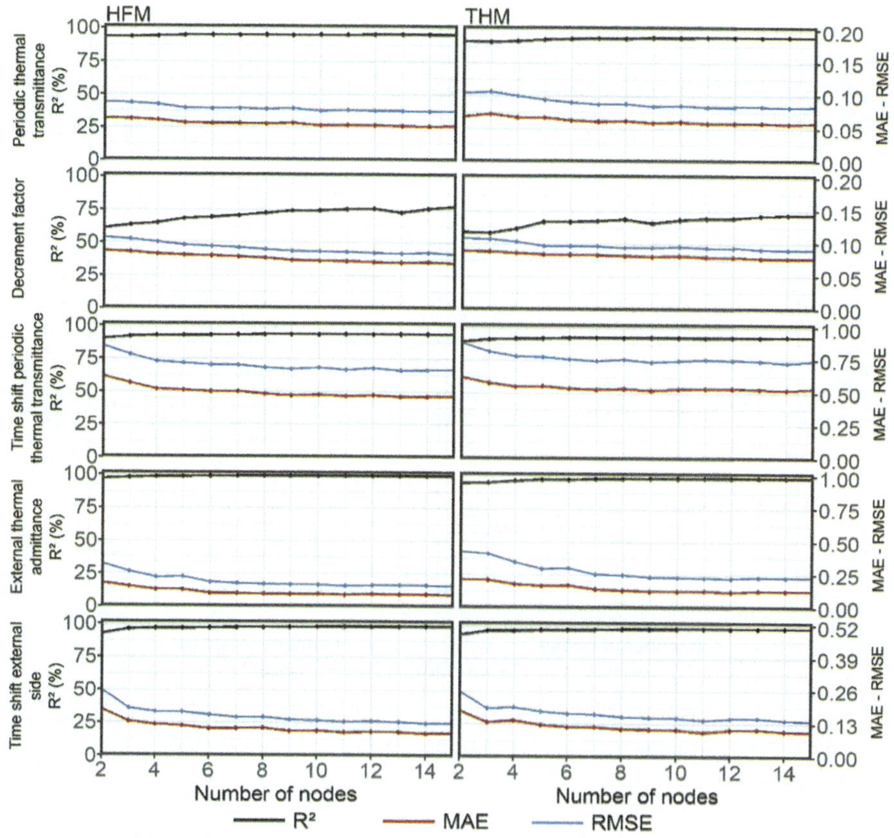

Fig. 5.1 Evaluation of R^2, MAE, and $RMSE$ in the training phase of the MLP models (the periodic thermal transmittance, the decrement factor, the time shift periodic thermal transmittance, the external thermal admittance, and the time shift external side)

varying the number of nodes of the hidden layer in the training phase with the HFM and THM approaches. As can be seen, the performances varied according to the type of variable analysed. Thus, there were variables with acceptable statistical parameters, whereas others obtained poor results in the training. In this regard, the periodic thermal transmittance, the time shift periodic thermal transmittance, the external admittance, the time shift external admittance and the external periodic thermal capacity obtained values greater than 90% in the determination coefficient, and the internal periodic thermal variables (the internal admittance and the internal periodic thermal capacity) obtained poor values in both the determination coefficient and error parameters. The MLPs of the decrement factor were in an intermediate position between the best and worst performances. In these models, the determination coefficient was greater than 70%. The optimal architecture of each model oscillated between 11 and 15 neurons in the hidden layer according to each thermal variable.

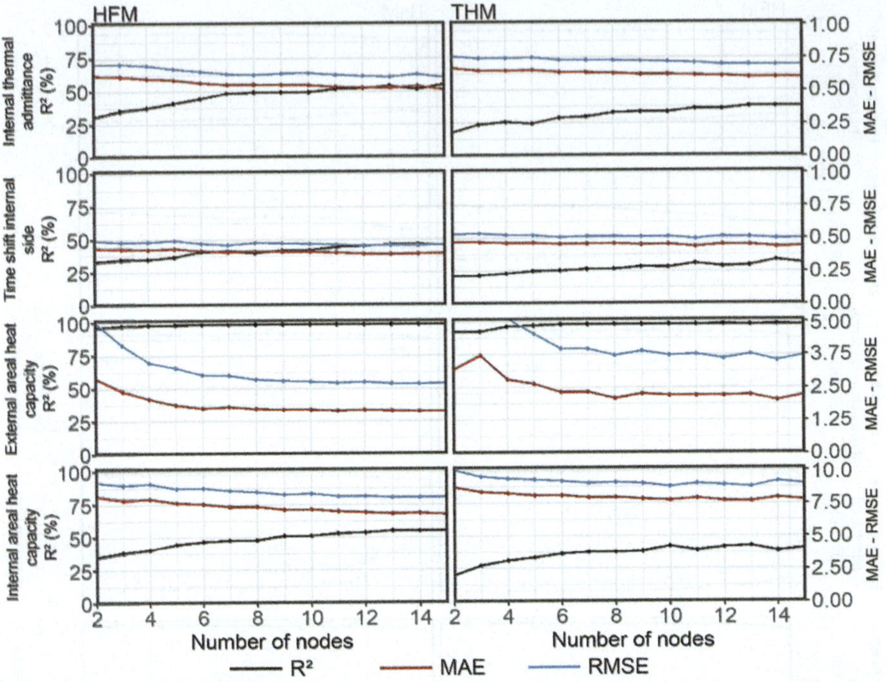

Fig. 5.2 Evaluation of R^2, MAE, and $RMSE$ in the training phase of the MLP models (the internal thermal admittance, the time shift internal side, the external areal heat capacity, and the internal areal heat capacity)

These optimal configurations were used to analyse the performance of the MLPs with the testing dataset. Table 5.3 shows the results obtained in the testing. The performances obtained in the testing phase followed the same tendency as in the training phase: the periodic thermal variables with better performances in the training obtained a determination coefficient greater than 90%, and the periodic thermal variables with worse performance obtained worse values in the determination coefficient. Nonetheless, the values obtained in the testing were better than those obtained in the training in some of the variables with worse performance. By way of an example, the determination coefficient of the variable of the internal thermal admittance went from 53.98% in the training to 60.61% in the testing. Likewise, the decrement factor was again in an intermediate position, with determination coefficients greater than 75%. A remarkable aspect of the results was the best performance obtained in the training and testing of the external periodic thermal variables. The use of indoor measurement variables could be the reason why the external periodic thermal variables obtained better estimations, although this aspect could not be accurately specified. The limitations related to the placing of the surface probes from the exterior (the heat flux and the surface temperature) and the test procedures of HFM and THM implied to consider only variables measured from the interior. In addition, another aspect found

Table 5.3 Values obtained in the statistical parameters in the testing phase with the optimal configuration of the MLP models

Variable	HFM			THM		
	R^2 (%)	MAE	$RMSE$	R^2 (%)	MAE	$RMSE$
Y_{12}	96.62	0.0477	0.0691	96.02	0.0531	0.0757
f	80.46	0.0658	0.0779	75.26	0.0751	0.0862
φ	94.04	0.4463	0.6519	93.79	0.5067	0.6725
Y_{11}	60.61	0.4976	0.5679	47.23	0.5671	0.6294
φ_{11}	56.31	0.3596	0.4090	34.54	0.4304	0.4771
Y_{22}	99.49	0.0716	0.1334	98.88	0.1000	0.1998
φ_{22}	98.10	0.0871	0.1289	97.76	0.0912	0.1400
k_1	60.27	6.7024	7.7119	49.74	7.3093	8.3316
k_2	99.22	1.5214	2.3667	98.47	1.9103	3.3239

in the results was the difference obtained with the two monitoring approaches (HFM and THM). The results showed that the performances of HFM were better than those of THM. In this regard, in HFM, there was an average increase in the determination coefficient of 5.94%, and an average decrease of 0.1382 in MAE and 0.2022 in $RMSE$ in comparison with THM.

These aspects can also be observed by analysing the histograms of the error obtained when estimating each instance of the testing dataset (Figs. 5.3 and 5.4). Given the variation of the periodic thermal variables, the analysis was focused on the percentage deviation between the actual and the estimated value. The percentage deviations obtained in the MLPs showed that regarding the external periodic thermal variables, there was a greater concentration of instances with percentage deviation values close to 0%, and regarding the internal variables, there was a larger number of instances that presented deviations greater than 10%. The histograms obtained with the periodic thermal transmittance are also emphasized, with low observation values in the error intervals, thus showing the impossibility of making estimations successfully with a low error in this variable. In general terms, the MLPs could successfully estimate the periodic thermal variables. This aspect can be seen in the results from the walls individually analysed (Table 5.4). These walls obtained estimations lower than 10% in most periodic thermal variables, except the internal periodic thermal capacity that obtained a percentage deviation greater than 10% in the wall B with the THM approach.

5.2.2 Random Forest

Like with the MLPs, the first step with RF was determining the optimal combination of trees for each model. As mentioned in Chap. 3, the number of trees is the

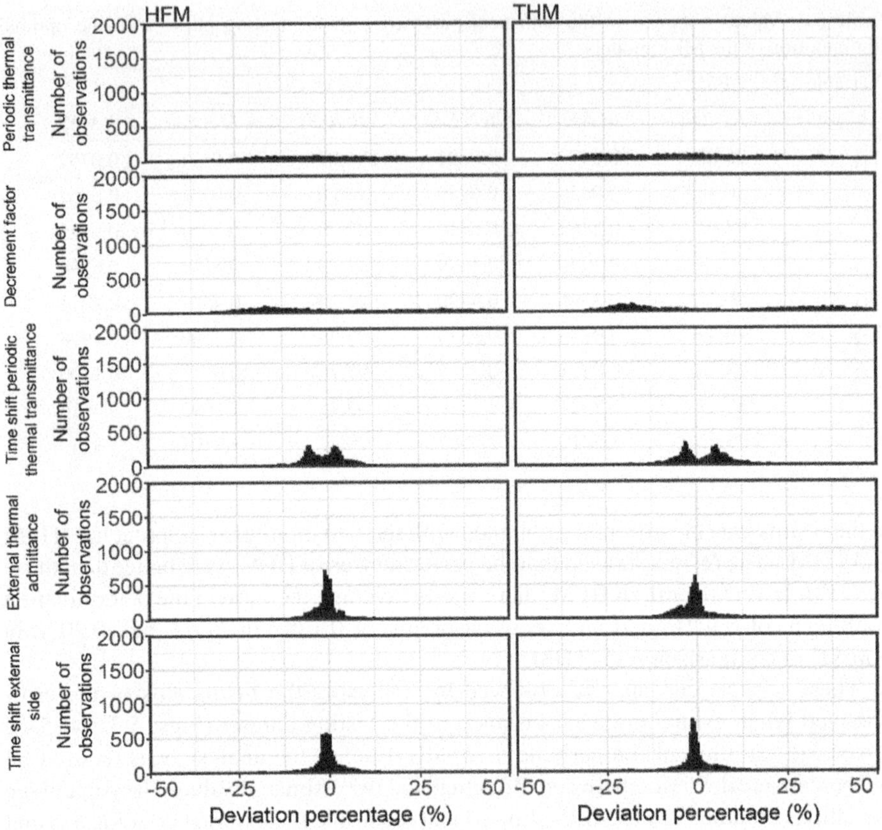

Fig. 5.3 Histograms with the deviation percentages obtained with the MLP models in the estimations of the periodic thermal properties of the instances of the testing dataset (the periodic thermal transmittance, the decrement factor, the time shift periodic thermal transmittance, the external thermal admittance, and the time shift external side). The histogram is represented with intervals of 1%

variable to be analysed in the design of these models. For this reason, configurations with numbers of trees between 2 and 50 were evaluated. Figures 5.5 and 5.6 show the values of determination coefficient and error parameters obtained in each model designed in the training phase. In general terms, the performance of the RF models was satisfactory: the determination coefficient was greater than 90% in all the combinations, and the error parameters obtained low values in comparison with the type of values of each variable. From models with a low number of trees to models with a larger number of trees, the performance was always appropriate. Nonetheless, the increase in the number of trees achieved better performances. In configurations with a number of trees between 40 and 45, the most optimal combination was achieved in each variable.

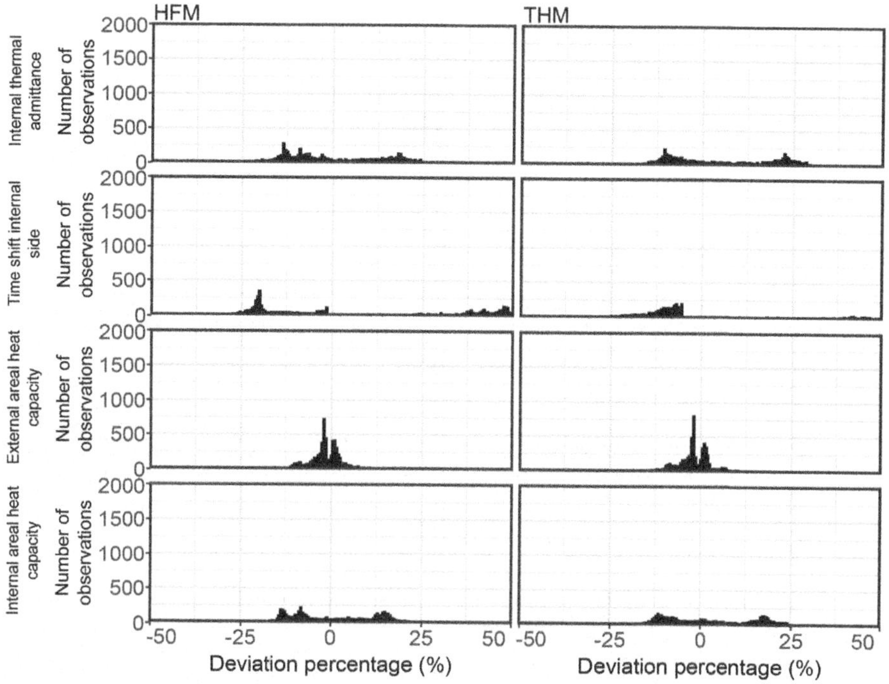

Fig. 5.4 Histograms with the deviation percentages obtained with the MLP models in the estimations of the periodic thermal properties of the instances of the testing dataset (the internal thermal admittance, the time shift internal side, the external areal heat capacity, and the internal areal heat capacity). The histogram is represented with intervals of 1%

Table 5.4 Error obtained in the estimation of the periodic thermal properties with the MLP models in the three walls individually analysed. The results are shown in absolute value

Variable	Percentage deviation (%)					
	HFM			THM		
	A	B	C	A	B	C
Y_{12}	3.01	5.97	5.52	7.30	4.51	6.00
f	5.60	8.21	5.55	5.27	8.62	7.10
φ	1.76	2.01	1.36	2.55	1.42	1.67
Y_{11}	7.71	5.03	6.67	7.03	6.68	9.79
φ_{11}	5.37	7.02	4.21	5.44	6.02	5.83
Y_{22}	1.67	2.79	1.83	2.25	2.68	2.34
φ_{22}	1.60	1.92	1.63	1.37	2.00	2.30
k_1	6.99	7.67	8.31	9.11	10.34	9.98
k_2	1.55	1.60	1.90	1.84	1.36	2.21

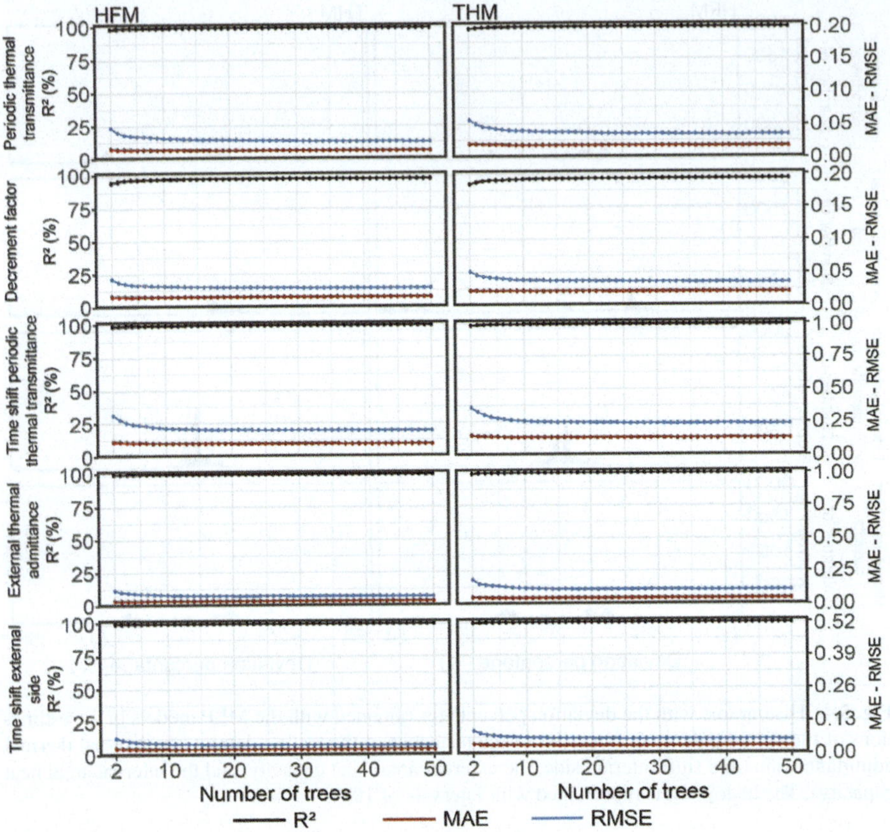

Fig. 5.5 Evaluation of R^2, MAE, and $RMSE$ in the training phase of the RF models (the periodic thermal transmittance, the decrement factor, the time shift periodic thermal transmittance, the external thermal admittance, and the time shift external side)

These optimal combinations were used to analyse the performance obtained in the testing phase (Table 5.5). The performances obtained with the two approaches were satisfactory in each variable. Regarding the HFM approach, the determination coefficient was always greater than 95%, and in the THM approach, it was greater than 93%. Likewise, the error parameters obtained low values, as can be seen in the case of the periodic thermal transmittance, with MAE of 0.0107 and 0.0153, or in the external thermal admittance, with MAE of 0.0301 and 0.0384. Regarding the performances obtained in the testing phase, there were two remarkable aspects in the performances of the MLPs: (i) first, although the performances were appropriate in all variables, there was a gradual tendency in the performance of the variables. Regarding the external thermal variables (e.g. the external thermal admittance), R^2 was greater than 99%, then the decrement factor was in an intermediate position (with R^2 of 96.65% (THM) or 97.89% (HFM)), and in the last position were the internal thermal variables (e.g. the internal thermal admittance) with the worst performances;

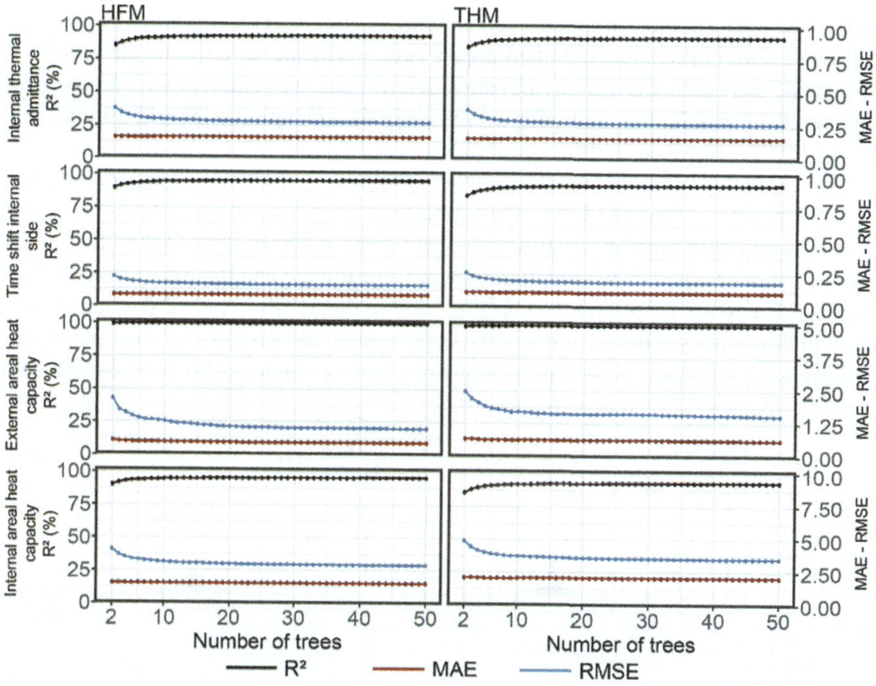

Fig. 5.6 Evaluation of R^2, MAE, and $RMSE$ in the training phase of the RF models (the internal thermal admittance, the time shift internal side, the external areal heat capacity, and the internal areal heat capacity)

Table 5.5 Values obtained in the statistical parameters in the testing phase with the optimal configuration of the RF models

Variable	HFM			THM		
	R^2 (%)	MAE	$RMSE$	R^2 (%)	MAE	$RMSE$
Y_{12}	99.68	0.0107	0.0214	99.36	0.0153	0.0302
f	97.89	0.014	0.0272	96.65	0.0195	0.0341
φ	99.47	0.0895	0.1959	99.20	0.1222	0.2419
Y_{11}	95.54	0.1079	0.2153	93.59	0.147	0.2582
φ_{11}	95.39	0.0777	0.1521	93.28	0.1076	0.184
Y_{22}	99.93	0.0301	0.0503	99.80	0.0384	0.0856
φ_{22}	99.81	0.0208	0.0415	99.63	0.0289	0.0571
k_1	95.63	1.432	2.8094	93.82	1.9433	3.3481
k_2	99.92	0.3811	0.7526	99.77	0.542	1.2815

and (ii) the performance obtained with HFM was better than that obtained with THM. Although both approaches obtained appropriate performances, the results obtained with HFM were characterized by having a determination coefficient with an increase between 0.13 and 2.11%, and with decreases in MAE between 0.0045 and 0.5113, and in $RMSE$ between 0.0069 and 0.5387.

These results are more easily seen by comparing the histograms included in Figs. 5.7 and 5.8. Many models made estimations with a percentage deviation lower than 10% and with high observation concentrations in errors close to 0%. This is the case of variables such as the external thermal admittance and the external areal heat capacity. Likewise, regarding the variables that in the MLPs obtained flat histograms and with low values in the error intervals, in the case of the RF models, greater concentrations were obtained in low percentage deviations, such as the periodic thermal

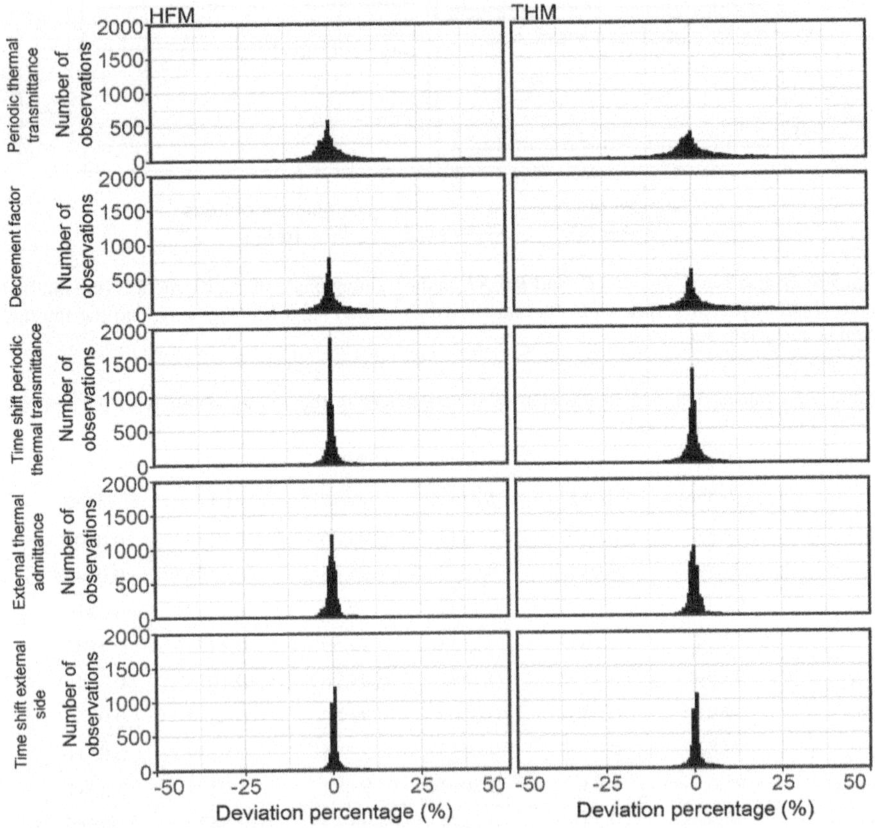

Fig. 5.7 Histograms with the deviation percentages obtained with the RF models in the estimations of the periodic thermal properties of the instances of the testing dataset (the periodic thermal transmittance, the decrement factor, the time shift periodic thermal transmittance, the external thermal admittance, and the time shift external side). The histogram is represented with intervals of 1%

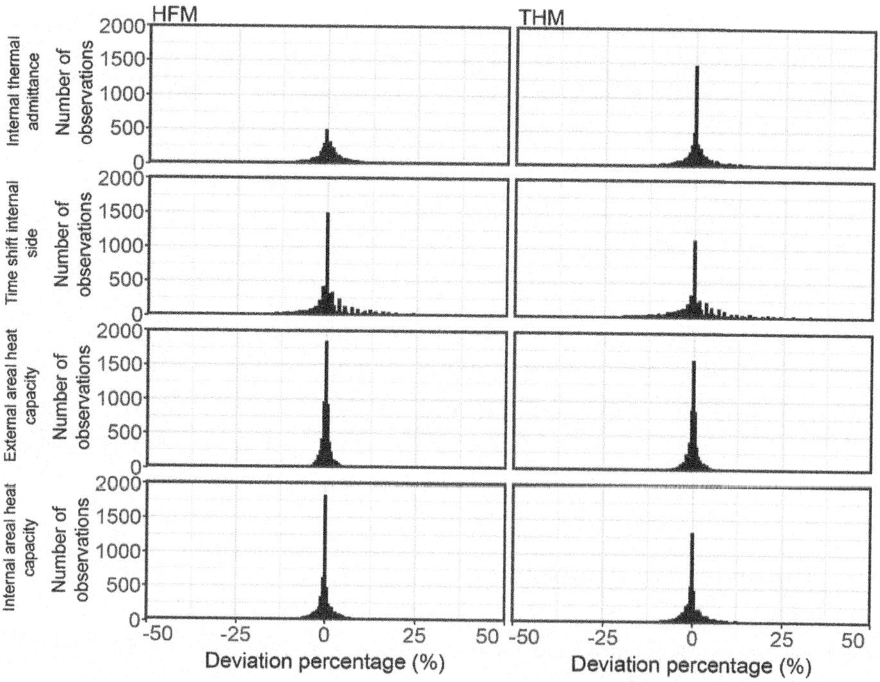

Fig. 5.8 Histograms with the deviation percentages obtained with the RF models in the estimations of the periodic thermal properties of the instances of the testing dataset (the internal thermal admittance, the time shift internal side, the external areal heat capacity, and the internal areal heat capacity). The histogram is represented with intervals of 1%

transmittance. However, not all the observations of the testing dataset obtained low percentage deviations as, sometimes, deviations greater than 25% were obtained. The reason was the values of some variables, such as the decrement factor, with values close to 0 (e.g. 0.097), and little variations of these values implied significant variations (e.g. for the value of 0.097, an estimated value of 0.1455 corresponded to a deviation of 50%). Although these percentage deviations could be significant, the individual analysis of these errors implies that the performances obtained by the models in these observations are acceptable. Nonetheless, these errors were obtained in a low number of observations, so it is generally expected that the performance of the RF models obtains low percentage deviations. This aspect can be seen in the results obtained in the walls individually analysed (Table 5.6). In the estimations made in the periodic thermal variables of these walls, most combinations of variables and wall obtained deviations lower than 2%, both with HFM and THM. However, deviations greater than 2% were sometimes obtained, such as the time shift internal side in wall B. Nonetheless, the maximum percentage deviation values (6.93% in HFM and 5.81% in THM) were low and acceptable to consider that the estimation made by the models was accurate.

Table 5.6 Error obtained in the estimation of the periodic thermal properties with the RF models in the three walls individually analysed. The results are shown in absolute value

Variable	Percentage deviation (%)					
	HFM			THM		
	A	B	C	A	B	C
Y_{12}	0.97	3.68	2.26	4.21	0.07	2.67
f	0.40	2.90	0.18	2.88	4.40	0.18
φ	0.05	0.83	0.22	1.17	0.33	0.49
Y_{11}	1.34	1.30	0.09	1.86	2.12	2.12
φ_{11}	1.31	6.83	1.77	1.23	5.81	1.77
Y_{22}	0.42	2.10	0.55	0.99	1.95	0.97
φ_{22}	0.00	0.35	0.59	0.17	0.92	1.18
k_1	0.92	2.92	0.87	3.00	2.77	2.85
k_2	0.14	0.23	0.82	0.57	0.01	1.40

5.3 Comparative Analysis

The possibility of estimating the periodic thermal variables using the MLP and RF models has been analysed. However, the performances detected by the two algorithms were different. Although these algorithms obtained good performance, RFs made better estimations than MLPs. This difference has been reflected in the aspects analysed (the performance in the training and testing, the error histograms, and the results in the walls individually analysed). One of the most significant aspects in the differences between the two algorithms is the concentration of observations with low percentage deviations: in the MLP models, the histograms had a low concentration of observations in deviations close to 0% and a greater distribution in high deviations, and in the RF models, there were always concentrations close to 0%. This aspect was not reflected in Chap. 4, where the results of stationary thermal transmittance obtained with the two algorithms obtained observations whose estimations had low error values. Although the performances varied according to the type of periodic thermal variable, RF obtained better performances than MLP in all assumptions. Thus, RF models are the most appropriate data analysis methodology to characterize the periodic thermal properties of the existing walls.

Another aspect is the differences between HFM and THM. These two approaches obtained very similar estimations and percentage deviations, but HFM was related to estimations with a lower percentage deviation. Nonetheless, the results obtained by the two approaches guarantees the reliability of using both; however, if it is possible to choose the experimental method when monitoring a wall, HFM should be given a greater priority than THM.

Chapter 6
Analysing with Artificial Intelligence Other Approaches to Experimental Thermal Characterization in the Existing Buildings

6.1 Introduction

This chapter aims to develop mathematical models based on MLP and RF that optimize the experimental methods to characterize the thermal transmittance. In particular, the models developed address the possibility of eliminating the error related to the theoretical formulation of THM. For this purpose, the thermal transmittance obtained with ISO 9869-1 (in which a theoretical value is not used in its formulation) is estimated by the input variables of THM. In addition, the building period of the wall should be known, as it is a requirement of the models developed for this approach. Although the building period could be determined through cadastral data, it is possible that these cadastral data are not useful to characterize the building period appropriately. In view of this situation, a new developed mathematical model would allow the building period of the wall to be known through monitored data.

6.2 Elimination of Errors in the Thermometric Method with Multilayer Perceptrons

As indicated in Chap. 2, one of the main weaknesses of THM is related to the use of a theoretical value for the total heat transfer coefficient. This theoretical value stems from ISO 6946 for a certain internal thermal comfort conditions and representative emissivity values of most construction materials. However, there could be differences between the actual heat flux of a building and the heat flux obtained by THM. Thus, the application of the method of ISO 9869-1 is related to a greater representation of results when the theoretical value of ISO 6946 presents limitations to be calculated.

To avoid these deviations, the results of the method of ISO 9869-1 were estimated through the variables and the procedures of THM. For this purpose, the successful

D. Bienvenido-Huertas and C. Rubio-Bellido, *Optimization of the Characterization of the Thermal Properties of the Building Envelope*, SpringerBriefs in Applied Sciences and Technology, https://doi.org/10.1007/978-3-030-63629-6_6

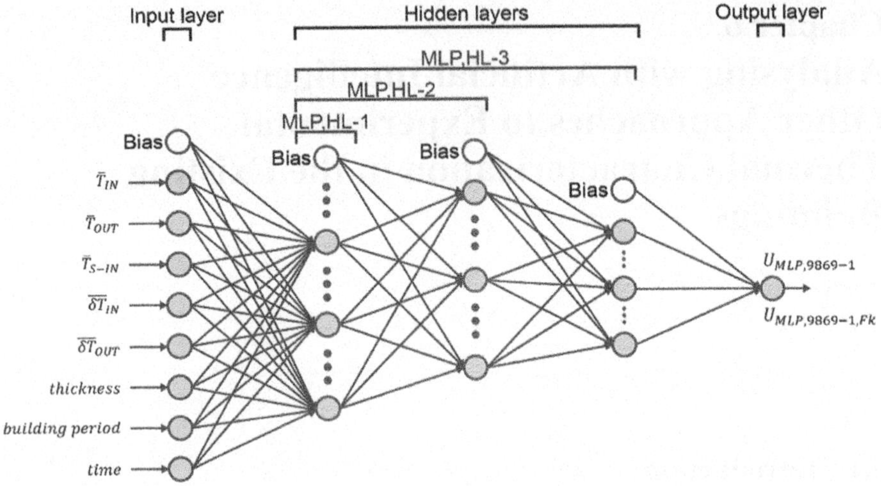

Fig. 6.1 Scheme of the MLP model considered to reduce the error in the results of THM

experience from the research on heat storage correction was the basis, using a significantly greater study sampling. The variables resulting from the method of ISO 9869-1 are the average method and the average method with heat storage corrections. The analysis was performed with MLP and the dataset designed in Chap. 3.

Likewise, the design of the variables of the MLP models was modified to encompass a wide variety of tests. For this reason, the same type of input variables of the previous research studies was used (i.e. indoor, outdoor, and surface indoor temperatures), as well as other variables required to characterize the thermal transmittance of ISO 9869-1 (both average and average with corrections for heat storage), including the difference of the average of temperatures of the first 24 h, the thickness, the time, and the building period (Fig. 6.1). The differences of the average of temperatures of the first 24 h were used to estimate the two typologies of output variables. Although these temperature differences are more related to the application of the heat storage correction procedures, the intention was that the structure of the two types of MLP was homogeneous to simplify the actual applications of the methodology.

In addition, the number of hidden layers was changed, studying architectures of 1, 2, and 3 hidden layers with a number of nodes that oscillated between 2 and 15. For the training of the models, 75% of the 22,820 simulated tests was used, and the remaining 25% was used for the testing. In addition, 12 wall typologies were individually analysed.

By analysing the results, the architecture with the best performance was first determined. In general terms, the models presented appropriate values for the determination coefficient and the optimal error parameters in the two approaches. Nevertheless, the most complex architectures slightly improved the statistical parameters, so the estimations were more adjusted in comparison with the actual values of the two approaches (Fig. 6.2). In this regard, the determination coefficient went from

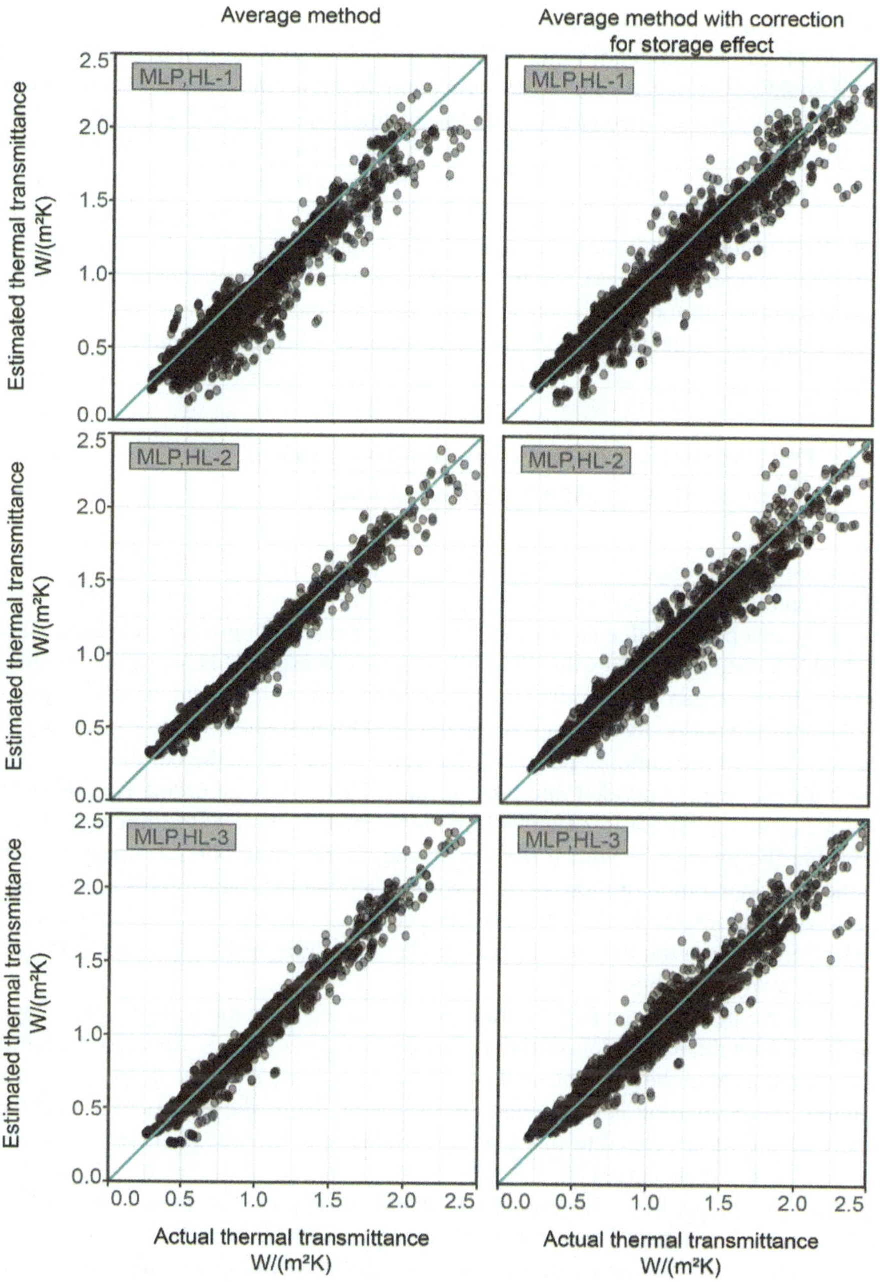

Fig. 6.2 Dispersion diagrams between the actual and the estimated values by each MLP model

98.30% in the MLP of one hidden layer to 99.19% in the model of two hidden layers and 98.81% in the model of three hidden layers. Thus, the thermal transmittance of HFM could be estimated with the test procedure of THM, thus eliminating the errors related to the use of a theoretical value for the total heat transfer coefficient.

The analysis continued with the influence of the building period and the possibility of using MLP models designed for each period based on the influence results obtained without knowing the input variables. One variable that presented a greater influence, together with the pure measurement variables (i.e. indoor, outdoor, and surface indoor temperatures), was the building period. Thus, not knowing the building period reduced the number of representative results in the walls individually analysed: from 100% when that period was known to 16.67% when it was unknown.

6.3 Determination of the Constructive Period of the Building with Monitored Data

Knowing the building period of the wall was therefore crucial for the previous procedure. Although this variable could be easily determined with the cadastral data of the building, it is possible that partial modifications are carried out in that building since its building date could imply that the wall does not correspond to the original building period. In this regard, the modifications carried out in buildings are usually adapted to the building period that belongs to the rehabilitation, so these modifications would be an update of the building period of the wall. As for Spain, the building period of the building stock is divided into three groups: P1 (before the normative NBE-CT-79), P2 (after the NBE-CT-79 and before the Spanish Technical Building Code), and P3 (after the Spanish Technical Building Code). Thus, an important aspect to study is the characterization of these periods without having the cadastral documentation. Following the line of work of the previous research studies, this building period is characterized with the data obtained from the monitoring with HFM and THM and the data analysis with RF.

Thus, the performance of the estimations obtained from the building period was analysed with the RF models of the dataset designed in Chap. 3. The input variables were those used in the research studies of Chapters 4 and 5. Unlike in other research studies, the estimation is of a qualitative variable. Thus, the statistical parameters analysed are those related to this type of variable (e.g. the true positive ratio or the area under the ROC curve).

After analysing the results obtained with the two approaches, the effectiveness of the RF models was seen to estimate the building periods of the wall. The results were satisfactory in the true positive rate and in the probability of estimating the building period correctly. The true positive rate obtained by the RF models oscillated between 99 and 100% (Fig. 6.3). Likewise, the building period was correctly estimated with the RF models. In this regard, the value of the kappa statistic was 0.998 in HFM and

HFM

	BP-1	BP-2	BP-3
BP-1	100.00%	0.00%	0.00%
BP-2	0.06%	99.72%	0.22%
BP-3	0.00%	0.08%	99.92%

THM

	BP-1	BP-2	BP-3
BP-1	99.71%	0.29%	0.00%
BP-2	0.28%	99.56%	0.17%
BP-3	0.00%	0.40%	99.60%

Fig. 6.3 Confusion matrices with the true positive rate values obtained with the two approaches

0.994 in THM, whereas the values of the area under the ROC curve were always 1 in all approaches and building periods.

Therefore, the results showed the possibilities to estimate the building period to automate and eliminate errors in the thermal characterization.

At this point, it is worth emphasizing the results obtained in Chap. 4 to characterize the thermal transmittance of ISO 6946. The use of this thermal transmittance value was crucial to validate the experimental results obtained with ISO 9869-1 or THM. Thus, a workflow based on data analysis models with MLP or RF to automate and optimize the characterization processes of the thermal transmittance in existing buildings could be established. The RF models could be used as an auxiliary evaluation methodology of the thermal transmittance before characterizing the thermal transmittance of the experimental methods. With the reference value obtained through the RF models, it could be seen whether the experimental results are coincident. If there is a deviation greater than 20% between both values, then the test is performed under inappropriate conditions or there is a type of error in the measurement conducted by the probes. Regarding the application of THM, the use of MLP would eliminate the errors related to the use of a theoretical value for the total heat transfer coefficient, using previously the RF model to determine the building period.

Finally, and to provide the reader with a visualization of the coupling of the optimization and automation methodologies developed, Fig. 6.4 includes a workflow using the two models. This workflow automates the whole analysis process for the thermal characterization of walls, so the possibilities of error of each calculation are eliminated. The workflow represented in Fig. 6.4 corresponds to an application of the models for THM. Regarding HFM, RF would be only applied to estimate the thermal transmittance of ISO 6946. Likewise, and although it is not included in the workflow, the RF models designed in Chap. 5 could be used to characterize the periodic thermal properties of buildings (although it is not included in the characterization process of ISO 9869-1 or THM). This workflow could be useful for architects, energy auditors, and engineers to characterize the thermal transmittance of walls in the existing buildings. In addition, the results are useful so that these technicians could have new

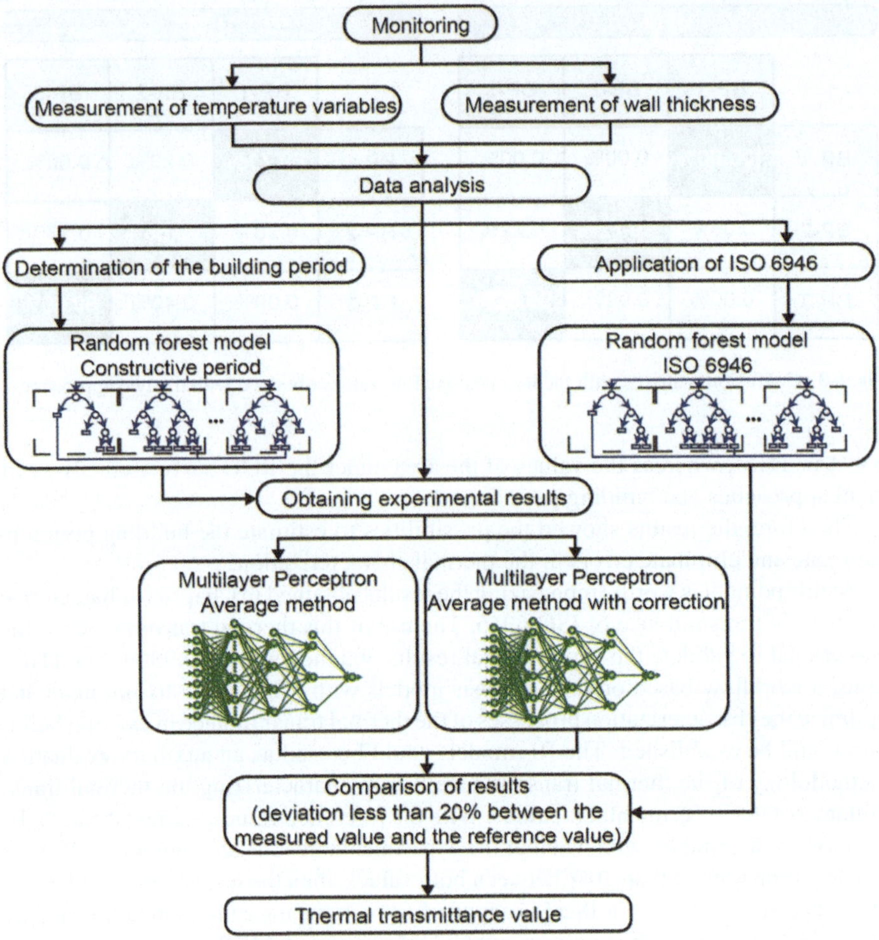

Fig. 6.4 Application flow of the automatic learning models developed to apply the experimental methods to characterize the stationary thermal transmittance

resources to guarantee the most appropriate proposal of energy saving measures by correctly characterizing the thermal properties of the existing building envelope. As a result, the energy renovation rate of the building stock would be higher, and the international goals to reduce greenhouse gas emissions by mid-twenty-first century would be achieved.

The manufacturer's authorised representative in the EU is Springer Nature Customer Service Centre GmbH, Europaplatz 3, 69115 Heidelberg, Germany. If you have any concerns regarding our products, please contact ProductSafety@springernature.com

Printed and bound by CPI Group (UK) Ltd, Croydon, CR0 4YY
30/09/2026
02225467-0002